BELNAP 55.00

DESERT SOIL FAUNA

John A. Wallwork, D.Sc.

Westfield College
University of London

PRAEGER

PRAEGER SPECIAL STUDIES • PRAEGER SCIENTIFIC

Library of Congress Cataloging in Publication Data

Wallwork, John Anthony, 1932—
   Desert soil fauna.

   Bibliography: p.
   Includes index.
   1. Desert fauna. 2. Soil fauna. I. Title.
QL116.W34   574.5′26404   81-12030
ISBN 0-03-055306-7   AACR2

Published in 1982 by Praeger Publishers
CBS Educational and Professional Publishing
a Division of CBS Inc.
521 Fifth Avenue, New York, New York 10175 U.S.A.

© 1982 by Praeger Publishers

All rights reserved

23456789 145 987654321

Printed in the United States of America

To Colleen,
in loving memory

Preface

It is generally considered that desert ecosystems are relatively simple in structure and functioning. Species diversity is often low, compared with cool, moist temperate ecosystems, and the factors that regulate ecosystem processes—energy flow and nutrient availability—are physical rather than biotic. Biological responses to extremes of temperature and moisture are usually clear-cut and more amenable to analysis than is the case in temperate grassland and forest. To one such as myself, who finds satisfaction in simplicity rather than complexity, deserts present an irresistible fascination. Moreover, since biological activity tends to be telescoped in the direction of the soil in hot deserts, soil biologists can claim a special interest.

This book is an attempt to cater to this interest; while there have been other books written about desert animals, I would venture to hope that the approach adopted here is original. By focusing on the desert soil system, it has been possible to run the gamut of invertebrate and vertebrate groups that form an integral part of the structure of these ecosystems and that contribute, in no small measure, to their functioning. Even so, the coverage is by no means uniform, and there are, perhaps, two main reasons for this. First, we know a good deal more about some groups of soil-based desert animals than about others. For example, there is a plethora of studies on the structure of ant and rodent "communities" in arid environments, while virtually nothing is known about the equally diverse and widespread soil microarthropods and nematodes. Second, some measure of selectivity had to be employed to prevent this book from getting out of hand. In exercising this selection I have chosen as my criterion the degree of dependence on the soil. Admittedly, this is a subjective, sometimes arbitrary, criterion, but it has allowed me to concentrate on those groups of animals whose activities impinge directly on the ecological events occurring within the soil, rather than on those that are merely located on its surface. As a consequence scant attention has been paid to birds and snakes, and precious little more to lizards. I make no apology for this, particularly since there is already in existence a voluminous literature relating to these groups.

After an introductory chapter that attempts to set the desert scene, the book falls more or less naturally into three main sections. Chapters 2, 3, and 4 are essentially concerned with the descriptive biology of the various groups of desert soil fauna. The middle section of the book (Chapters 5, 6, and 7) deals with the nature of the driving variables, moisture and temperature, and the biological response to them. In the last three chapters an attempt is made to identify common themes, convergences, and parallelisms that characterize the life-styles of desert soil animals.

A colleague recently remarked, "I suppose that, in writing a book of this kind, you rely for about eighty percent of your material on the work of others; only twenty percent is your own stuff." In the present case this is an overgenerous estimate of my contribution to desert biology, but it does give an idea of the immense scope of desert research, and it underlines the need for a multidisciplinary approach. In truth I have drawn heavily on the literature, particularly on symposium volumes produced under the auspices of the Zoological Society of London and the American Institute of Biological Sciences; in addition, on the reports issued as part of the International Biological Programme Desert Biome studies. I have also been fortunate in having had access to material in manuscripts not yet published and would like to acknowledge the generosity of Walter G. Whitford (New Mexico State University), Clifford Crawford (University of New Mexico), Penny and John Greenslade (South Australian Museum and Commonwealth Scientific and Industrial Research Organization, Adelaide), Perseu F. Santos (University of Rio Claro, Brazil), Diana Freckman (University of California-Riverside), and Yossi Steinberger (New Mexico State University). Unpublished information has also been volunteered by a number of my colleagues, in addition to those already mentioned, and I would like to acknowledge the assistance provided by James F. McBrayer (Oak Ridge National Laboratory), Dac Crossley (University of Georgia), Jo A. Springett (New Zealand Ministry of Agriculture and Food), Mark A. Dimmitt (Sonoran Desert Museum), Peter F. Newell (University of London), and Michael F. Brown (New Mexico State University).

Several colleagues have provided photographs for use in this work and acknowledgments have been made in the appropriate places. This is, however, hardly a sufficient tribute to their unstinting generosity. In particular I would like to express my sincere thanks to John L. Cloudsley-Thompson (University of London), Diana Freckman, Richard E. MacMillen (University of California-Irvine), Walter G. Whitford, Clifford S. Crawford, Mark A. Dimmitt, and Terry A. Vaughan (Northern Arizona University) for coming so readily to my aid. The efforts of Gerhard Ott (Pomona College), Chris Walker, and Haidee Price-Thomas (both of Westfield College) in the photographic department are also greatly appreciated. Thanks are also due to Bernardette Kamill (British Museum of Natural History, London), the Trustees of the British Museum (of Natural History), Macmillan, and Springer Verlag for permissions to include published material.

At the outset of this project I considered it imperative to obtain a genuine "feel" for the desert and its fauna. This I have been able to do largely through the good offices of one man: Walter G. Whitford, who has been my guiding genius over the past two years. Many of the ideas developed in this book owe their origins to discussions conducted in conference halls and seminar rooms and out on the bajada; here also, I would like to acknowledge the considerable contributions made by Clifford S. Crawford, Diana Freckman, Perseu F. Santos, and Yossi Steinberger. I owe a special debt of gratitude to John Gurnell (University of London) for his interest in this project, for drawing my attention to literature I would otherwise have overlooked, and for many helpful discussions. Also to Susan Badger (Eau Claire, Wisconsin) for her excellent work in copyediting the manuscript, I express my gratitude.

Most important, though, the friendship and hospitality shown to me by Walt and Linda Whitford are the most cherished rewards to accrue from this work. This book was written for them. They will understand why I have chosen to dedicate it otherwise.

Contents

Preface vi

1 Desert Types and Topography 1

Deserts of the World 1
The Biological Factor 7
Desert Topography 10

2 The Surface-Active Fauna: Invertebrates 28

Centipedes 29
Arachnids 31
Millipedes 39
Woodlice 45
Snails 51
Insects 53

3 The Surface-Active Fauna: Vertebrates 65

Marsupials 65
Amphibians 67
Reptiles 69
Rodents 72
Lagomorphs 82

4 The Subterranean Fauna 84

Nematodes 85
Enchytraeids 88
Myriapods 91
Collembola 93
Diplura 96
Insects 96
Arachnids 100
Mammals 112

5 The Rainfall Factor and the Biological Response 115

Physical Effects 115

ix

	Biological Effects	116
	The Biological Response	119
	Drought Avoidance	123
6	**Drought Tolerance**	132
	Water Gain and Water Conservation	132
	Surface Predators	134
	Surface Detritivores and Herbivores	144
	The Subterranean Fauna	165
7	**The Temperature Factor and the Biological Response**	169
	Global Patterns	170
	Local Patterns	171
	Seasonal Patterns	172
	Diurnal Patterns	173
	The Biological Response	175
8	**Species Diversity and Resource Allocation**	189
	Biotic Associations	189
	The Ecological Species	192
	Species Diversity	192
	Resource Allocation: The Subterranean Fauna	193
	Resource Allocation: The Surface-Active Fauna	198
9	**Life-Styles for Survival: Tactics**	216
	Morphological Traits	218
	Behavioral Traits	223
	Physiological Adaptations	236
10	**Life-Styles for Survival: Strategies**	245
	Body Size	246
	Environmental Predictability	248
	Response to the Environment	249
	Reproductive Rate and Longevity	251
	Population Changes	254
	Conclusion	260
References		262
Name Index		283
Subject Index		287

1
Desert Types and Topography

The overriding characteristic of deserts is their aridity. More precisely, deserts are regions on the earth's surface where annual evapotranspiration exceeds precipitation; where water loss by direct evaporation from the soil and by transpiration from plants is greater than water input from rain or dew formation, on a yearly basis. The rate of evapotranspiration is a function of temperature, and it follows that deserts form under different rainfall regimes in different parts of the world. In the subtropics, for example, a desert will form where the annual rainfall is 120 mm. In cooler parts of the temperate regions a desert will form only under conditions of lower rainfall because the rate of evapotranspiration is lower. In polar regions precipitation in the form of rain is negligible, at least at extreme latitudes, and desert conditions prevail here. Polar deserts do not fall within the scope of this review, but the remainder do, and they are listed in Table 1.1. Their distribution on a worldwide basis is shown in Fig. 1.1.

DESERTS OF THE WORLD

The boundaries drawn around the deserts depicted in Fig. 1.1 are approximations. Their precise locations depend on who is doing the drawing and what criteria are used to define the limits. This subjectivity arises because although it is generally recognized that deserts are arid places, aridity is expressed to different degrees in

TABLE 1.1.
The Major Deserts of the World

Desert	Area (km² × 10⁶)	Cool	Hot	Inland	Coastal
Sahara	9.1	−	+	+	−
Australian	3.4	−	+	+	−
Arabian	2.6	−	+	+	−
Turkestan	1.9	+	−	+	−
North American	1.3	+	+	+	+
Patagonian	0.67	+	−	+	+
Thar	0.60	−	+	+	+
Kalahari/Namib	0.57	+	+	+	+
Takla Makan/Gobi	0.52	+	−	+	−
Iranian	0.39	+	−	+	−
Atacama	0.36	+	−	−	+

Source: Compiled by the author from various sources.

different deserts. A distinction can be made between arid and semiarid zones, for example, and some authorities would consider only the former to qualify as true deserts. Viewed in this light, deserts are climatic phenomena and, as such, have been variously classified by Thornthwaite (1948), Meigs (1953), Köppen (1954), and Walter and Stadelmann (1974), among others. On the other hand, it is important to bear in mind that desert boundaries are not fixed and unchangeable. Zones of aridity expand and contract in the long term not only in response to climatic changes but also to biotic influences (see below). This dynamic character provides a basis for distinguishing between actual and potential desert areas (Dregne 1970; Kassas 1970). It also introduces the idea that biological criteria may be incorporated into the definition of a desert system. For example, Whittaker (1970) suggested that deserts could be separated from other terrestrial ecosystems by virtue of their low productivity. MacMahon (1979) used plant and animal distributions to define the limits of the various North American deserts. We could dwell at length on these and other definitions of deserts that appear in the literature, but this would serve no good purpose since this exercise has already been carried out by McGinnies (1979a, 1979b). As far as this present volume is concerned, deserts are loosely defined as areas where the soil fauna, actually or potentially, is exposed to environmental stress; animals that are so exposed are not particularly concerned as to whether they live in deserts or semideserts.

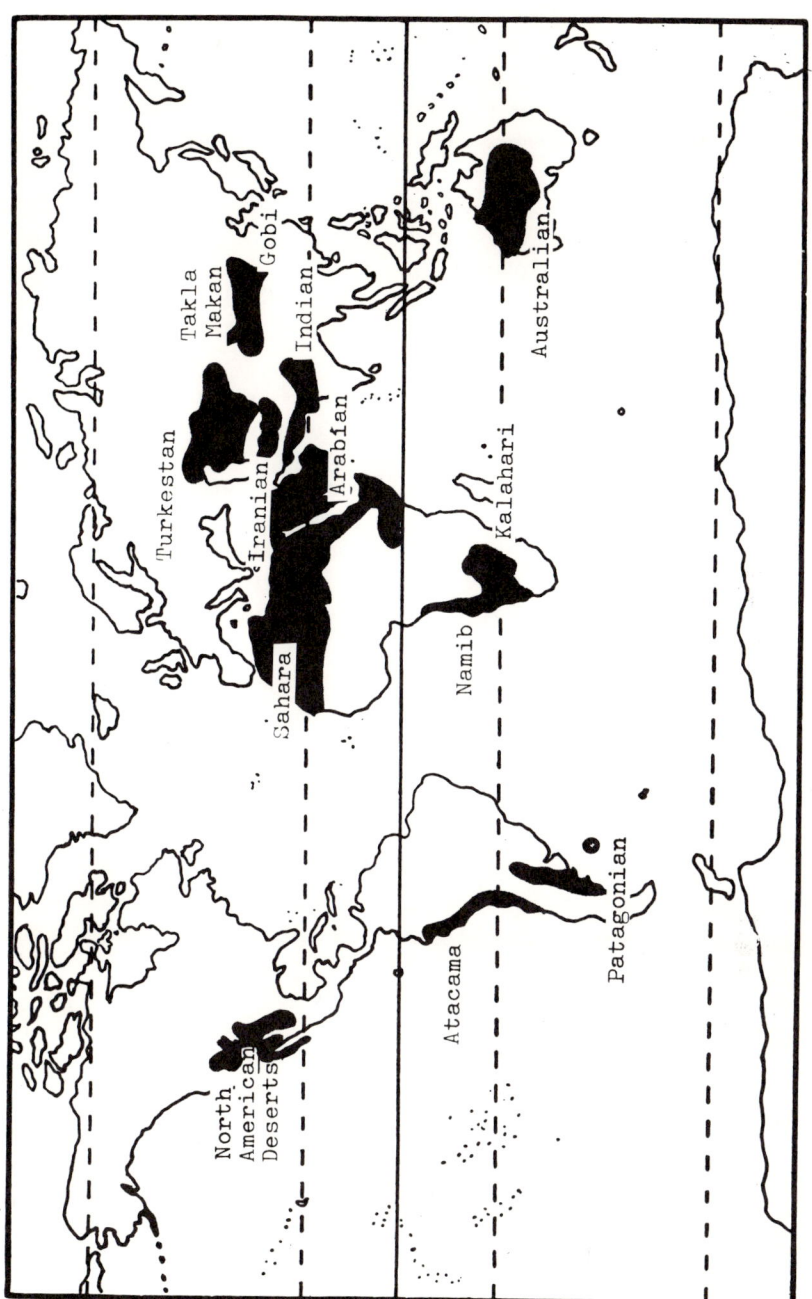

Fig. 1.1. World distribution of deserts.

Returning now to the arid areas delineated in Fig. 1.1, these can be broadly divided into two major categories: hot and cool. Hot deserts are distributed globally across the tropics of Cancer and Capricorn, north and south of the equator. To this category belong the Saharan and Kalahari deserts of Africa, the Arabian desert, the Great Australian desert, the Thar desert of India, and the deserts of the southwestern parts of the United States and northern Mexico (Figs. 1.2-1.5). Cool deserts may have average summer temperatures as high as those in hot deserts, but average winter temperatures are much lower. This category includes the Great Basin desert of Nevada, the Turkestan and Takla Makan/Gobi complex of Asia, and the Atacama and Patagonian deserts of South America.

The creation of hot subtropical deserts is commonly attributed to the large-scale movements of dry air from the equatorial region, which, in turn, are the result of the earth's rotational motion. These air masses, heated over the equator, rise by convection to high altitudes where they are cooled, lose their moisture, and are deflected north or south. Over the subtropics these cold, dry air masses descend and, warmed by compressional, adiabatic heating, pick up moisture from the land and its vegetation. This warmed,

Fig. 1.2. Sonoran desert dominated by ocotillo (*Fouquieria splendens*) in the northern part of Baja California, Mexico.

Fig. 1.3. Cholla "garden" (*Opuntia bigelovii*) in Joshua Tree National Monument, Mojave desert, California.

Fig. 1.4. Smoke tree (*Dalea spinosa*) in the Colorado desert near Glamis, California.

Fig. 1.5. High Sonoran desert near Tucson, Arizona, dominated by the saguaro cactus (*Cereus giganteus*).

moist air then rises and, as it cools in the upper atmosphere, discharges its moisture as rain over the temperate regions. Subsequently, it is further deflected toward the Poles and here descends to repeat the cycle.

These global weather patterns are reinforced by other more localized climatic effects to produce deserts. Coastal deserts, such as the Atacama of Peru and Chile and the Namib of southern Africa, are formed by exposure to winds that have been cooled by passing over cold ocean currents. These cooled winds may be ladened with moisture, but it rarely condenses to form rain. The deserts so formed fall into the category of "cool" rather than "hot." Cool deserts also occur in the interior of continents in temperate latitudes where the sheer remoteness from the oceans is coupled with relatively low temperature regimes. These two factors combine to ensure that the winds blowing over these interior deserts have lost much of the moisture they contained when they originated over the oceans. Examples of these deserts are the Turkestan and Takla Makan/Gobi deserts of central Asia and the Great Basin desert of Nevada and Utah. The latter is also influenced by rain shadow effects. Rain shadow deserts are formed to the leeward of mountain ranges, and these, in the case of the Great Basin, are the Sierra Nevada and Cascade mountains to the west. These ranges cause prevailing westerly winds from the Pacific to drop their moisture on the western slopes as they are deflected upward into cooler regions of the atmosphere. The air currents that descend on the eastern slopes are dry, and they pick up moisture from the land. Other examples of deserts that are influenced by rain shadow effects are the Mojave and the Patagonian desert.

THE BIOLOGICAL FACTOR

Climate plays an important role in determining where deserts develop, but other nonphysical factors may come into play to maintain and, indeed, to extend the range of existing deserts. These factors are biological and are associated with the activities of man and other animals.

Human Effects

Man's activities, particularly those concerned with agriculture and animal husbandry, have an important impact on the development of deserts. This is becoming increasingly a problem in the southwestern parts of the United States, for example. Here cattle ranching and the cultivation of cotton require irrigation schemes. This may present no problems in the regions adjacent to the Colorado, Rio Grande, and Pecos river systems, but in more remote areas water has to be pumped up from underground lakes (Fig. 1.6). This "fossil" desert water is a nonrenewable resource for two reasons. Present-day rainfall levels are insufficient to balance the deficit produced by the demands of irrigation, and, perhaps more seriously, the subterranean caverns from which this irrigation water is drawn tend to collapse. These collapsed reservoirs could not refill even if water were available.

Additionally, there is the problem created by tilling the soil and substituting monocultures of crops, which are harvested periodically, for a varied, natural vegetation growing in undisturbed soil (Fig. 1.7). This agricultural practice loosens the soil, renders it more friable, and prevents the accumulation of a protective surface layer of litter. As a consequence soil erosion occurs and desert conditions are promoted.

Overgrazing by domesticated animals is a potent factor in maintaining deserts and contributing to their extension. In recent decades this problem has become particularly acute in northern

8   Desert Soil Fauna

Fig. 1.6. Tapping "fossil" water on a cattle ranch near Portal, Arizona.

regions of Africa bordering the Sahara. People in these regions husband cattle, sheep, and goats as a main source of protein and, by deep-seated tradition, allow free-range grazing, as opposed to a more controlled regime, to fatten their livestock. Not surprisingly, grazing pressure along the desert margins is intense, resulting in the removal of surface vegetation and the exposure of soil to evaporative water loss and erosion (Kassas 1970).

Other Animals

Grazing pressure on desert systems is also exerted by herbivorous groups of the native fauna. Locusts in the Sahara and lagomorphs in the Chihuahuan desert can be cited as examples of desert animals that remove considerable quantities of the surface vegetation. Again, seed-feeding is a "key industry" in hot deserts and is practiced, in the main, by three groups of animals—ants, rodents, and birds (see Chapters 2, 6, and 8). Seeds represent an important reserve compartment in the desert ecosystem model, and the extensive exploitation of this reserve by granivores will have repercussions on any subsequent pulse of vegetational activity.

Fig. 1.7. Cotton monculture in the Sonoran desert of southern Arizona.

However, this is a two-edged sword since caching of seeds just below the soil surface, and their subsequent abandonment by rodents, may facilitate germination (Soholt 1973; Reichman 1979).

Ants and rodents may properly be considered as members of the soil fauna since, in hot deserts at least, they have a subterranean base. It is more appropriate to include them in this review than it is to treat locusts, rabbits, and birds, which are much less dependent on the soil. However, we must not lose sight of the fact that the feeding activities of, for example, lagomorphs impinge on the events occurring in the soil/litter subsystem (see Chapter 3). There is a reciprocity between ants and rodents as far as their granivorous activities are concerned, and this will be examined in more detail later. For the moment we are only concerned with drawing attention to the fact that these feeding activities may have an impact on already well-established desert systems. They do not create deserts—they merely contribute to their perpetuation. On the other hand, recent studies by Valiachmedov (1981) have identified a situation in which a group of soil animals actually creates areas of desert by virtue of their presence and activities. These animals are termites, *Anacanthotermes ahngerianus*, and the desert areas with which they are associated are known as takyrs.

Takyrs are alkali soils occurring in the cool Turkestan desert.

They are reminiscent of gray semidesert soils, or sierozems. Organic content is low, and they have a high content of exchangeable sodium, a compacted crust, and low biological activity. The takyr region is stepped and terraced, sloping to rivers, with dune formations stabilized by low vegetation and divided by interdune valleys that are devoid of vegetation except during the early spring when algal crusts form on the soil surface. These interdune valleys are the provenance of *A. ahngerianus*, and they are located around the periphery of the takyrs. The true takyr is not colonized by the termites, but detailed examination of takyr soils indicates that they were formed initially in sites where the termites were active. The insects abandoned these sites, possibly because of overexploitation, possibly because of changes in microclimate. Whatever the reason, termite nests collapsed and filled, and the stratigraphy of takyr soils is consistent with this in-filling hypothesis.

DESERT TOPOGRAPHY

Hot deserts vary in their ability to support biological communities. We have only to contrast the bare expanse of Saharan sand dunes with the more richly vegetated parts of the California Mojave to appreciate this fact. On a much more localized scale marked vegetational differences also occur within particular desert sites. For example, in the area around Alamogordo, New Mexico, there is an expanse of gypsum sand dunes colonized very sporadically by mesquite (*Prosopis*), iodinebush (*Suaeda*), sand sage (*Artemisia filifolia*), and the gypsophilic *Muhlenbergia villiflora, Dicranocarpus parviflorus, Nama carnosum, Dalea filiciformis,* and *Coldenia hispidissima* (MacMahon 1979). (See Figs. 1.8, 1.21, and, 1.22.) In immediate proximity to these dunes there are dense stands of saltbush (*Atriplex*) (Fig. 1.9) providing extensive ground cover over salt flats. These low-lying saline soils probably develop as a result of successive temporary floodings; they are collection points for surface runoff from neighboring mountains during the rainy season and, as such, will accumulate salts as the floodwater evaporates. In many areas this is a natural process—the consequence of a topography in which high mountains are dissected by wide, sloping alluvial valleys—and it gives rise to salt flats in which only halophilic plants such as *Atriplex* are able to survive.

This example illustrates the fact that the topography of the desert surface has an important influence on its hydrological characteristics and, through these, on the nature and distribution

Desert Types and Topography 11

Fig. 1.8. Gypsum dunes at White Sands National Monument, New Mexico. Vegetation in the foreground is sand sage (*Artemisia filifolia*).

Fig. 1.9. Saltbush desert, White Sands, New Mexico.

of the vegetation. This vegetation grows along the courses followed by surface water runoff (see Chapter 5) and around the margins of temporary lakes. Of even greater importance, as far as the soil biologist is concerned, is recognition of the fact that this surface runoff redistributes organic litter: the base for detritus food chains in all soil systems. This redistribution is very much a function of desert topography. In developing this idea further, particularly in relation to patterns of biological activity in hot deserts, it is necessary to devise some scheme for the classification of soil habitats. Surface relief and soil type are, perhaps, the primary variables that need to be incorporated into such a scheme. Others, such as the stability of the substratum and patterns of vegetation and of organic litter accumulation, follow from these.

Surface Relief

In an early study of a Mojave soil ecosystem, a distinction was drawn between high plateau and low desert (Wallwork 1972a, 1972b). The former extends northward from the San Bernardino and Chocolate mountain ranges of southern California into eastern California and southern Nevada. This plateau desert supports a relatively rich vegetation consisting of the Joshua tree (*Yucca brevifolia*), piñon pine, juniper, smoke tree (*Dalea spinosa*), and Mojave yucca (*Yucca schidigera*) and a variety of low shrubs such as catclaw acacia (*Acacia greggii*), senna (*Cassia armata*), *Krameria*, and *Artemisia*, along with various cacti (Fig. 1.10). The lowland desert, extending southward from Palm Springs to Yuma, is often referred to as the Colorado desert, although many authorities consider it as a westward extension of the Sonoran. With the latter this low desert supports extensive stands of creosote bush (*Larrea tridentata*) and burr sage (*Ambrosia* sp.) (Fig. 1.11).

This distinction between high and low desert is an oversimplification. It does not take into account other topographical variables present in hot deserts. In the Chihuahuan desert, for example, the principal feature of the landscape is the gently sloping valley (bajada), as much as 60 km across, flanked by precipitous mountains that may be flattopped (mesas). These montane regions are rock deserts, subject to severe erosion by wind and water. The products of this erosion (gravel, sand, clay) are deposited on the sides and floor of the bajada variously to form gravel aprons, claypans, and sand dunes, which support a range of vegetation types (Fig. 1.12). The coarse soils of the high bajada support

Desert Types and Topography 13

Fig. 1.10 High plateau desert, Joshua Tree National Monument, California, with Joshua tree (*Yucca brevifolia*) and *Juniperus* sp.

Fig. 1.11 Low sagebush desert near Palm Springs, California.

14 Desert Soil Fauna

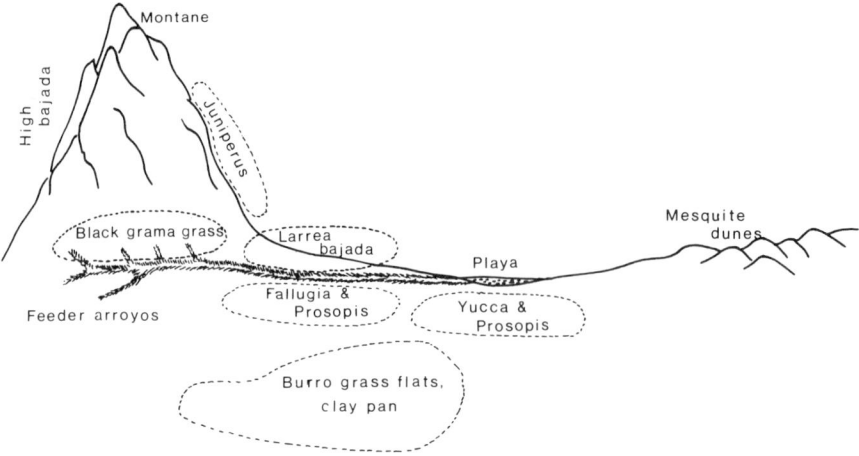

Fig. 1.12 Schematic representation of Chihuahuan desert topography. (Suggested by W. G. Whitford.)

scattered growths of desert oak (*Quercus grisea*), juniper (*Juniperus monosperma*), sotol (*Dasylirion wheeleri*) mesquite (*Prosopis juliflora*), Apache plume (*Fallugia paradoxa*), and various yuccas, notably the soaptree yucca (*Y. elata*) (Fig. 1.13); in sheltered depressions the ground may have a cover of black grama grass

Fig. 1.13. Soaptree yucca (*Y. elata*).

(*Bouteloua eriopoda*). At lower elevations the vegetational characteristics of the bajada change as the larger perennials give way to smaller growth forms, such as creosote bush, snakeweed (*Guterriezia sarothre*), and pepperweed (*Lepidium lasiocarpa*) (Fig. 1.14). Similar patterns of zonation also occur in the Sonoran desert (MacMahon 1979) where the giant saguaro (*Cereus giganteus*) and paloverde (*Cercidium* sp.) plant communities of the high bajada contrast with the creosote bush/bursage assemblages lower down (Figs. 1.5 and 1.11).

This pattern of zonation is complicated by the development of arroyos: steep-sided outwash channels carved into the bajada by rainwater runoff. These arroyos represent corridors along which plants (and animals) of the high bajada can extend their range into lower elevations. This is particularly well seen in the foothills of the Chiricahua Mountains of southeastern Arizona, where juniper trees spread out into the alluvial plain along the margins of dry washes. The low point of this alluvial plain (playa) is the site of a temporary lake during the rainy season. Playa sites have a moderate to high salinity and, as a consequence, are rather sterile from a biological point of view. However, a characteristic type of vegetation develops around their margins with Tabosa grass, *Hilaria mutica*, often prominent.

Topographical features such as bajadas, arroyos, and playas

Fig. 1.14. Snakeweed (*Guterriezia sarothre*).

16 Desert Soil Fauna

Fig. 1.15. Rocky scree on high bajada, southern Arizona.

(Figs. 1.15-18) not only influence the hydrology of desert regions but, in turn, are influenced by this. Intense rains move rock fragments from place to place, erase existing arroyos, and create new ones such that the surface relief of the desert is a dynamic rather than a static system. At the same time desert topography has a certain measure of universality, and the mesa/bajada/arroyo/playa occurs under different guises in geographically distinct desert regions. In the Sahara, for example, the montane rock desert is termed the hammada, the rocky aprons of the high bajada are scree slopes, while the more gentle slopes of the lower bajada find their African counterpart in the serir or reg (gravel plains). Again, the arroyos of North American deserts are the wadis of the Sahara; in the latter, the sand dune field is termed an erg (or ereg), and fine-textured saline flats (playas) are variously designated dayas, sebkhas, chotts, or kavirs (see McGinnies 1979a).

Wind action is also a potent factor in shaping the desert landscape. High winds are characteristic of hot deserts. Air currents move mineral particles from place to place and have an abrasive effect on exposed rock formations, causing their physical disintegration. Clay and silt particles can be moved great distances and eventually deposited as loess. Sand particles often accumulate around deep-rooted shrubs, such as mesquite, which may offer

Desert Types and Topography 17

Fig. 1.16. Saltbush (*Atriplex* sp.) on low bajada, near Alamogordo, New Mexico.

Fig. 1.17. Arroyo near Casa Grande, Arizona, flanked by ironwood and apache plume.

18 Desert Soil Fauna

Fig. 1.18. Playa lake after rains, near White Sands, New Mexico.

some measure of stability, sometimes leading to dune formation. Extensive dune systems, oftentimes attaining a height of 200 m (Cloudsley-Thompson 1975b), devoid of vegetation are features of the Sahara, Namib, Turkestan, Takla Makan/Gobi, and Arabian deserts, but they occur only sporadically in other deserts (Seely 1978). Localized dune formations, such as those that occur in North American deserts, are often threatened by overgrazing and human incursions (Figs. 1.19 and 1.20), which reduce their stability. The impact of these activities on the desert landscape is probably quite severe, on a local scale, but the effect on the soil fauna has received little attention.

Like their counterparts in cool, moist temperate coastal regions, desert dunes can be classified into mobile, semipermanent, and permanent formations (Figs. 1.21 and 1.22). This classification places emphasis on stability, but, despite its importance in the ecological scheme, it has been given little attention in relation to the soil fauna. As an alternative, dune formations have been classified according to their shape and mode of formation. In this way a distinction can be made between crescentic-shaped barchan dunes, formed by the action of moderate winds on relatively small volumes of sand, and seif dunes, which develop as elongated ridges composed of large volumes of sand fashioned by strong winds

Desert Types and Topography   19

Fig. 1.19. Human impact statement I. Sand dunes near Glamis, California.

operating along the length of the dunes and also at right angles to them. Seif dunes are examples of longitudinal dunes; they are elongated in the direction of the prevailing winds. Barchan dunes, on the other hand, present a broad leading edge to a unidirectional wind that is deflected laterally to produce the crescent shape. This is a variant of the transverse dune formation, which occurs as giant sand ripples created by the action of moderate, unidirectional winds on light sand.

Transverse dunes, barchans, and seifs are mobile in character, and they move in the direction of the prevailing winds. As we have seen, the establishment of vegetation is an important factor in stabilizing a dune—rendering it stationary and permanent. However, relatively stable dune formations can occur in the absence of vegetation in regions where the wind blows from all directions; the result is a star dune.

Dune topography is of particular interest to the desert soil biologist since it provides the key to resource distribution in sandy soils, preeminently for tenebrionid beetles. Martz (1980), for example, noted that *Cryptoglossa* sp. fed mainly on dead bees and the feces of the kangaroo rat *Dipodomys*, which collected on the leeward side of barchan dunes in the southern Mojave. This

20   Desert Soil Fauna

Fig. 1.20. Human impact statement II. Sand dunes at White Sands, New Mexico.

sheltered habitat contrasted with the more exposed slip face, inclined at a maximum angle of 35°, where food items were less plentiful; as a consequence beetles moved at greater speeds to increase the probability of encountering these items. Even more attention has been devoted to the definition of habitat types within the dune system of the Namib desert.

The Namib is a cool, coastal desert with extensive dunes, devoid of vegetation, disrupted by rocky mountains and infrequent bare gravel plains and riverbeds (Seely 1978). Edney (1971) identified dune slopes, crests, and slip faces as distinct habitats in a bare dune formation and contrasted these with dune bases and interdune valleys, which are colonized by grasses and herbs. These variations in surface relief provide a topographical diversity that offers a range of microhabitats for tenebrionid beetles. These coleopterans are particularly successful in the Namib, and the extent of their adaptive radiation can be judged from the fact that 35 genera and about 200 species have been recorded from this desert. In contrast other deserts of the world in which bare dune formations are less extensively developed support a much less diverse tenebrionid fauna (Seely 1978). This topic is explored in more detail in Chapter 8, but it may be noted in passing that the

Fig. 1.21. Semipermanent dunes, White Sands, New Mexico.

richness of the tenebrionid fauna in the Namib cannot be explained solely in terms of the presence of dune habitats; climate and geography are also important.

Desert Soils

Much has already been written about desert soils (see, for example, Brown 1968; Cloudsley-Thompson 1975b; McGinnies 1979b), and it would serve no good purpose to embark on an exhaustive account here. Instead, it is proposed to place emphasis on those features that figure prominently in the role of the desert soil as an environment for animal life. In this regard it is important to make an immediate distinction between soils that are essentially mineral in character and those that contain appreciable amounts of organic materials.

Mineral soils are much more widespread in deserts than are organic soils. The former, which are properly considered as embryonic or azonal, are relatively unstructured, and they either develop in situ, through the weathering of bedrock, or they develop after the transportation of mineral particles by wind and water. In situ weathering produces gravel plains (serir or reg in Africa, billy gibber in Australia) in low-lying areas and aprons of rock fragments (screes and bajadas) at higher elevations. In both of these

22 Desert Soil Fauna

Fig. 1.22. Mobile dunes, White Sands, New Mexico.

situations calcium carbonate concretions occur at or near the surface, forming a pavement or gravel known, in North America at least, as caliche. Transported soils are windborne deposits (loess) or waterborne alluvial deposits. These sediments can be distinguished, on the basis of particle size, into clay, silt, and sand categories. Because of their smaller particle size, clays and silts are often transported great distances by wind currents and flash floods and accumulate in basins and playas where alternate dry and wet conditions induce the formation of claypans. These pans experience interrupted drainage and support halophytic shrubs and other plants that can tolerate periodic anaerobic root environments. Sand particles are rolled or blown into dune formations, and, as already noted, these formations vary in their stability and permanence. They also vary in their chemical composition, and here a distinction can be made between silica and gypsum dunes. However, dune chemistry would seem to be less important than dune topography in influencing the distribution and diversity of the soil fauna, although, again, this is an area that has not attracted the attention of desert soil zoologists. In this regard the faunal characteristics of such desert soil types as the sierozem, solonchak, and solonetz merit further study.

The desert soil is not everywhere barren and inhospitable, and indeed, it has been calculated that the productivity of vegetation growing along the margins of arroyos approaches that of a deciduous forest (Ludwig and Smith 1978). The organic litter produced by this vegetation, in the form of leaves, wood, and seeds, forms the nutrient and energy base for desert soil communities and is vital to their existence. The extent to which this organic material mixes with the mineral soil varies considerably from place to place and, as we have seen, is a function of hydrology and topography. Organic substrates of appreciable thickness accumulate on the high bajada and along arroyo margins under large shrubs, such as mesquite, desert oak, Apache plume, juniper, paloverde, ironwood (*Olneya*), and desert willow (*Chilopsis*) (Figs. 1.17 and 1.23). These organic accumulations remain sharply distinct from the underlying mineral soil (Wallwork 1972a) and represent stable "island" environments for soil animals.

A number of different major soil conditions can be identified in hot deserts, and the summary presented in Fig. 1.24 is based on those occurring in the southwestern deserts of the United States. The scheme presented here does not accord with current pedological terminology, but it provides an environmental framework for a study of the distribution and diversity of desert soil fauna. A knowledge of the conditions for life offered by a given soil type allows us to assess, for example, how successfully it can be exploited by burrowing animals—and this may be a reflection of its stability. It also allows us to gain some idea of its amenability to litter accumulation and of the consequent effects of soil microclimate. Each of these parameters can now be considered.

Soil Stability and Burrowing Activity. It is axiomatic that rocky soils, screes, and gravels provide greater problems for burrowing animals than the more finely textured clays, silts, and sands. The ability to burrow in the soil is often a vital factor for survival in hot deserts. It is not surprising, perhaps, that most of the best-adapted members of the soil fauna, such as the desert cockroaches, harvester ants, granivorous rodents, millipedes, woodlice, and tenebrionid beetles, are found much more frequently on the fine, deposited soils than on rocky aprons and scree slopes. However, the development of claypans and caliche will restrict burrowing activities on alluvial plains and in interdune valleys; so it is not simply a question of differentiating between the rocky habitats of the high bajada and the finer transported soils of valley basin floors. At each of these topographical levels the soil provides a range of substrata, some more accessible to burrowing animals than others. Rocks and gravels in montane regions, with their innumerable

24 Desert Soil Fauna

Fig. 1.23. Litter accumulation and the shade effect under desert juniper in the Joshua Tree National Monument, Mojave desert, California.

cracks and crevices, offer refugia, sheltered microhabitats, notably for some of the larger-sized predatory invertebrates. In the deserts of the American Southwest such refugia are occupied by the

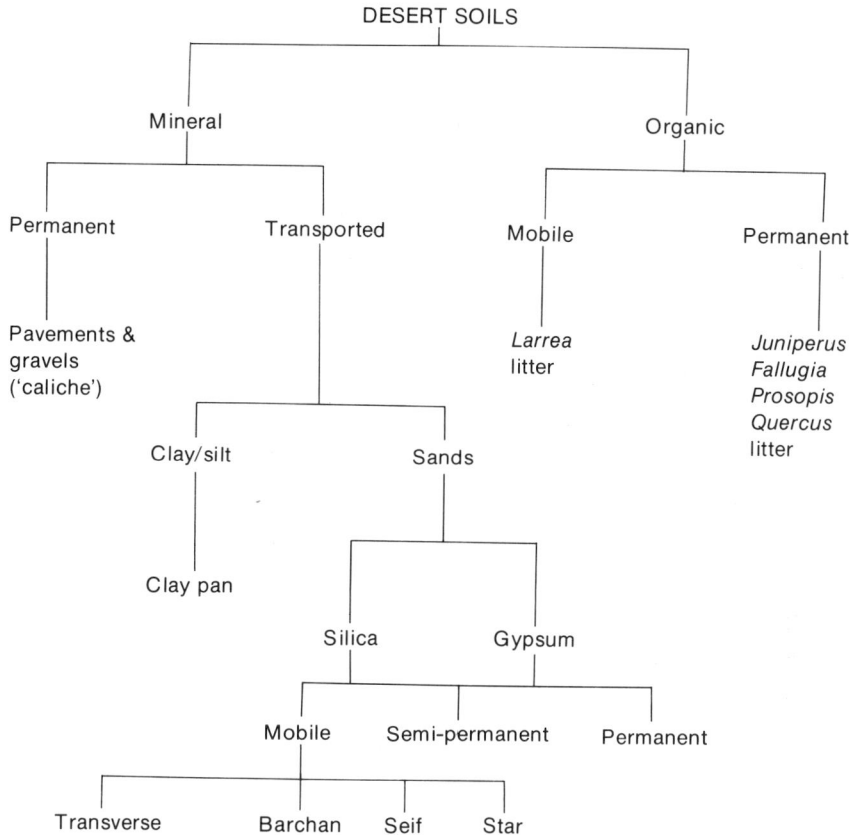

Fig. 1.24. Classification of desert soil types occurring in the Chihuahuan desert of the southwestern United States.

centipede *Scolopendra polymorpha* and the scorpions *Centruroides sculpturatus* and *Diplocentrus spitzeri*. *S. polymorpha* and *C. sculpturatus* cannot properly be considered as burrowing animals, but *D. spitzeri* excavates a shallow burrow beneath its rock shelter. Many scorpions are effective diggers, and this ability allows them to colonize some desert habitats that would offer a measure of resistance to other less well-equipped burrowing forms.

Amenability to Litter Accumulation. Litter accumulation on desert rock pavements is largely prevented by the erosional forces to which these surfaces are subjected. This statement needs some qualification, however, since at higher elevations litter accumulates in protected situations under trees and shrubs (see above). In the foothills of the Chiricahua Mountains of Arizona, for example, stable litter occurs under juniper growing through a surface pavement of caliche. The stability of this litter is enhanced by growths of fungal mats and other binding agents such as resinous exudates from the tree itself (Wallwork 1972a). In such sites a relatively rich fauna of soil microarthropods develops, and the community structure is more diversified than in other parts of the desert.

Sand dunes are the most unstable of all desert soil types, and their impermanence is one of the factors that militates against their colonization by large desert shrubs. The organic input in such systems is obviously low, and the shifting character of the dunes will cause any litter that is produced to be dispersed widely. Animals that habitually occur in desert sands, such as the roach *Arenivaga* and tenebrionid beetles, subsist very largely on wind-blown organic particles (see the example of *Cryptoglossa* mentioned earlier).

Soil Microclimates. The soil types that we have been considering vary considerably in their microclimatic characteristics, in the stability of their moisture and temperature regimes, in the extremes of temperature and moisture that they experience, and, by these tokens, in their suitability as environments for soil-dwelling animals.

Extremes of temperature and moisture are more pronounced in desert soils than in perhaps any other kind of soils. With regard to temperature, attention is usually focused on the upper extreme since soil surface temperatures can rise to between 60°C and 70°C in the hottest deserts. However, it must be remembered that even the so-called hot deserts can experience subzero temperatures during the winter season. Animals living here have to combat the dual problems of overheating and freezing. Similarly, although deserts by definition are considered to be arid, the concentrated nature of the rainfall event, if and when it occurs, produces flooding, particularly in low-lying areas.

Rock pavements and gravels are less prone to flooding than are the playa soils, particularly where the scree and caliche are on the sloping valley sides. Rainwater percolates easily through stony rubble; pavements and caliche are impervious but, because of their position on the bajada, will tend to dry out quickly after rain since this water removes itself as surface runoff. Again, because of their elevation these rocky soils may have lower yearly maximum temperatures and lower yearly minima than low-lying alluvial soils. In their own way these embryonic soils present problems for survival as far as desert animals are concerned.

Sandy soils, particularly those that develop into dunes, are well drained, but moisture content can be higher here than on pavements and caliche, especially in the leaf catchment area of the sparse vegetation colonizing these dunes. This moisture supply will also be augmented by internal dew formation. Extremes of temperature will occur on the surface of sandy soils because of the paucity of an insulating cover of ground vegetation. However, the drainage properties of sandy soils will prevent them from becoming inundated during the rainy season.

It is the organic soils that offer the most suitable environments for animals, and it is here that the richest faunal diversity occurs (see Chapter 8). The organic material, consisting of leaf material, wood fragments, and fungal growths, provides a substantial food base for the saprophage trophic level, but it also provides a relatively stable microclimate. The fact that organic soils develop in shaded localities in hot deserts where temperatures may be as much as 10°C lower than in exposed sites (Wallwork 1972a) means that the amplitude of the circadian temperature curve is less than that in unshaded localities. Short-term variations in moisture content and relative humidity will be lowered since leaf litter and humus have greater water-holding capacities than mineral particles. The stability of the moisture regime is also enhanced by the relatively equable temperature characteristics of organic soils.

Of the various sites considered above consolidated organic substrates clearly provide the best conditions for the soil fauna to exercise its decomposer function. However, these permanent organic soils represent only very localized sites of net primary production. Walter Whitford (1979: personal communication) has calculated that about 70 percent of net primary production input to desert soils in the Chihuahuan and Sonoran regions is derived from creosote bush litter that is transported down the bajada and along the arroyos. Here, it is dispersed and mixed with clays and silts, and there is evidence (Santos, DePree, and Whitford 1978) that decomposition pathways are regulated by the interactions between mites that prey on nematodes feeding on the true decomposers, bacteria. But more of this later.

2
The Surface-Active Fauna: Invertebrates

The desert soil provides shelter for various vertebrates and invertebrates that emerge periodically to forage on its surface and in the aerial vegetation. Such activity in above-ground habitats hardly qualifies these animals as permanent members of the soil fauna if we define a true soil animal as one that spends the whole of its life within the soil/litter system. However, in hot deserts many of these surface foragers transport organic material from the surface to subterranean sites where decomposition may proceed more rapidly. Small mammals, harvester ants, and termites are important in this respect. There are herbivorous forms or, more specifically, granivores (gramnivores or seed-eaters), and they exist, cheek by jowl, with the permanent, subterranean fauna that is essentially detritivore based. Hitherto, no attempt has been made to integrate the activities of these two components of hot desert systems: the surface-active and subterranean elements. This is a challenge that falls within the compass of this book. In this chapter and the next we concentrate on the surface-active fauna, and although a distinction is made between invertebrates and vertebrates, as a matter of convenience, this is largely arbitrary.

The surface-active invertebrate fauna of hot deserts includes both primary and secondary consumers. The former are represented for the most part by millipedes, woodlice, mollusks, and insects. Predatory invertebrates active on the soil surface are centipedes and arachnids, and these are considered first.

CENTIPEDES

The centipedes of cool, moist temperate soils in Europe belong to the Lithobiomorpha. Their counterparts in warmer regions of the world are the Scutigeromorpha and Scolopendromorpha, with the latter predominating in hot deserts. Scolopendromorphs have a number of features that are particularly suited to life in arid environments. In the first instance desert scolopendromorphs tend to be much larger in body size than their cool, moist temperate relatives. This can be interpreted as a morphological adaptation that reduces the surface area/volume ratio and, as a consequence, the rate of water loss across the body surface. Second, scolopendromorphs are nocturnal.

Nocturnalism is general among centipedes the world over; so in this sense scolopendromorphs are preadapted to colonize environments where the relative humidity is higher at night than during the day. Cloudsley-Thompson and Crawford (1970) used aktographs to monitor the activity patterns of a species of *Scolopendra*, possibly *S. polymorpha* (Fig. 2.1), collected from the desert around Albuquerque, New Mexico, and discovered a nocturnal rhythm that persisted in continuous darkness but was almost lost in constant light. This suggests that *S. polymorpha* normally spends the daylight hours under rocks or in crevices and soil burrows. There is no real evidence that centipedes in general—and *S. polymorpha* in particular—actively produce burrow systems in the soil (see Crawford and Riddle 1974, for example). However, desert centipedes undoubtedly utilize crevices under rocks and probably also the abandoned burrows of scorpions and small mammals.

Scolopendromorphs of arid and semiarid environments remain something of an ecological enigma. Virtually nothing appears to be known about their feeding habits or their ability to regulate the densities of whatever they feed on. These predators, in turn, represent prey items for the burrowing owl *Speyotyto cunnicularia* (see Cloudsley-Thompson and Crawford 1970), but, again, the extent to which this activity regulates the population density of *Scolopendra* sp. is not known. According to Craybill (quoted in Cloudsley-Thompson and Crawford 1970), there are three, possibly four, closely related species of *Scolopendra* occurring in the deserts of southwestern United States. This fact raises a number of intriguing questions. Do two or more of these species exist sympatrically? Does habitat separation occur, and if so, what are the mechanisms that maintain this? To what extent is sympatry related to differential resource allocation? Essentially, all of these questions are

Fig. 2.1. The centipede *Scolopendra polymorpha* from the Pelocillo Mountains, New Mexico/Arizona border. (Photo courtesy of C. S. Crawford and reproduced by kind permission of Macmillan Press.)

concerned with the twin ideas of species packing and niche breadth—topics that for years have been agonized over by desert ecologists working with rather complicated systems, such as rodents and ants, for example. Scolopendromorphs in deserts provide a simpler system, worthy of further investigation.

Xeric and semixeric centipedes have more rapid growth rates and shorter life-spans than the lithobiomorphs of cooler, more moist regions. The latter have a life duration of two to three years or more, while *Scolopendra amazonica* completes its life cycle within a year in the savannah region of northern Nigeria (Lewis 1972). Similarly, the scolopendromorph *Rhysida nuda togoensis* and *Ethmostigmus trigonopodus* (also from northern Nigeria) may breed when one year old, which is unusual among centipedes. This may be regarded as an adaptation to life in a pulse environment where biological activity is discontinuous and survival places a premium on the rapid completion of postembryonic development. There are parallels here with members of the subterranean fauna (see Chapter 4.).

Maternal care is very much a feature of postembryonic development in centipedes. As far as scolopendromorphs are concerned, Lewis (1972) cites the studies of Brunhuber (1970) on *Cormocephalus anceps*, an inhabitant of mesic environments, which exhibits parental care for four months during a period of postembryonic development that spans at least two years. This species evidently has a longevity of at least six years. In contrast the more abbreviated periods of postembryonic development and longevity in the semiarid *Rhysida nuda togoensis* and *Ethmostigmus trigonopodus* would suggest a curtailment, or even absence, of brood care, although Lewis (1972) presents no concrete evidence one way or the other.

ARACHNIDS

Chelicerate arthropods are well represented in the surface-active fauna of hot deserts, notably by the scorpions, solifugids, uropygids, and spiders.

Scorpions

Desert scorpions belong, in the main, to four families: Buthidae (Fig. 2.2), Scorpionidae, Diplocentridae, and Vejovidae. The first three of these enjoy a wide distribution in the tropical and

32 Desert Soil Fauna

Fig. 2.2. The Sudan scorpion *Leiurus quinquestriatus*. (Photo courtesy of J. L. Cloudsley-Thompson.)

subtropical regions of the Old and New Worlds. Vejovids are the most common scorpions in North American deserts; to this group belongs *Hadrurus arizonensis*, one of the best adapted of all desert scorpions.

In many ways arid environments are the provenance of these large-sized arachnids and their relatives the whip scorpions, camel-spiders (solifugids), and scorpion-spiders, although this statement should not be taken to imply that these groups evolved in these environments. It is just as probable that they arose in mesic environments and subsequently extended their distribution into deserts. Scorpion species, for example, show differing degrees of tolerance to arid conditions, and this suggests that an invasion from mesic habitats has been mounted with greater success by some species than others. Like the centipedes, some desert scorpions live in crevices under stones; others excavate burrows in the soil. All are active during the hours of darkness.

When scorpions are active, they are foraging for food. Species that forage on the surface have an endogenous activity rhythm that causes them to emerge from their refugia at night. The North American *Centruroides sculpturatus* and *Paruroctonus mesaensis* and the African *Buthus hottentotta* are examples of this group

(Hadley and Williams 1968; Toye 1970; Crawford and Krehoff 1975). But there are species such as the North American *Diplocentrus spitzeri* (= *peloncillensis*) that remain in a subterranean habitat. Here there are no environmental clues to distinguish between day and night. Consequently, this species may forage in a more or less continuous manner and perhaps should be considered as a "permanent" soil inhabitant (see Chapter 4).

Surface-foraging scorpions differ in their feeding habits. Some species actively seek prey, whereas others, and *P. mesaensis* is an example, move only a short distance from the burrow entrance, preferring to sit and wait for the prey to come to them (Polis 1979; Polis and Farley 1979).

It is likely that scorpions, in general, will accept what is available in the way of prey—that is, they are opportunistic or facultative feeders. This is suggested by the work of Polis (1979) on the desert sand scorpion, *Paruroctonus mesaensis*, which is common in the dunes around Palm Springs, California. This species was observed to accept 95 prey species, of which 10 figured consistently. These regular items in the diet consisted of various tenebrionid beetles, the desert sand roach *Arenivaga investigata*, the harvester ant *Solenopsis xyloni*, gryllids, and other scorpions of the same and different species. It is of interest to note that *P. mesaensis* preyed on other members of its own species to the extent that such prey ranked fourth in importance of all the prey species taken. Clearly, cannibalism is a significant feature of the ecology of this desert scorpion, and probably of many others.

Scorpions are generally regarded as viviparous; the young may have a "placental" connection within the body of the mother where they undergo a gestation period of varying duration, depending on the species. After birth the young scorpions climb onto the back of the female who cares for them until they can cope for themselves. Comparative data on reproduction and development for a wide range of scorpions are given by Polis and Farley (1979), and these indicate that the desert vejovid *P. mesaensis* is very much an average species. It has a clutch size of about 33, compared with an average of all scorpions considered of 31; it has a gestation period of 10 to 14 months, compared with the overall 9; and its age at maturity varies between 19 and 24 months, which is close to the general average of about 27 months. However, if the comparison is restricted to species that inhabit arid regions (Table 2.1), some interesting differences emerge. *P. mesaensis* produces markedly fewer young, on average, than *Androctonus australis*, *Buthus occitanus*, and *Leiurus quinquestriatus*, and they are retained in the body of the female for longer periods of time. Despite the latter

TABLE 2.1.
Comparison of Reproductive and Development Patterns between Various Species of Desert Scorpions

Species	Age at Maturity (Months)	Gestation Period (Months)	Clutch Size
Androctonus australis	21.7–26	8	46
Buthus occitanus	6	9–11.5	50
Leiurus quinquestriatus	8	5	57
Paruroctonus mesaensis	19–24	10–14	33

Source: Adapted from Polis and Farley 1979.

fact, growth to maturity takes considerably longer than is the case with *B. occitanus* and *L. quinquestriatus*. A relatively long gestation period may be interpreted as a means of increasing the survival chances of the young in species, such as *P. mesaensis*, which live in extremely arid environments. Similarly, such increased prospects of survival would obviate the necessity for a large clutch size. The consequent saving in energy could be vital to survival in view of the unpredictability of food supplies in a sand desert. However, it is more difficult to interpret the slow growth to maturity shown by both *P. mesaensis* and *A. australis*. The former species undergoes six molts during postembryonic development, as does *B. occitanus*, while *A. australis* molts seven times. Moreover, looking at scorpions in general, there is no apparent relationship between the number of molts and the age at maturity. Most molt six or seven times, yet maturation times range from 6 months to more than 80 months. Growth and ecdysis are functions of food intake and temperature (Polis and Farley 1979) and, as such, vary seasonally. Slow development times may reflect the general food levels and climatic characteristics of sites colonized by *P. mesaensis* and *A. australis*. The fact that cannibalism is a major feeding activity of *P. mesaensis* may indicate the scarcity of other prey. Indeed, this habit is one of the factors that produces 60 percent mortality among the young before they attain the second instar in this species (Polis and Farley 1979).

Paruroctonus mesaensis, like other vejovids, mates in the

period from spring to autumn, and the young are born during a more restricted period (July to September). This synchronization of births probably results from an earlier pulse of prey, in the spring, and gives rise to synchronous surface activity of second instar juveniles in early August. Polis and Farley (1979) observed this activity to occur at this time in each of five consecutive years. Restricted periods of recruitment also occur in other desert arachnids, notably the soil-dwelling mites (see Chapter 4).

Solifugids

Variously called camel-spiders, sun-spiders, or wind-scorpions, solifugids occur in all of the world's deserts except the Australian. They are highly efficient surface predators by virtue of their ability to run very rapidly over the ground surface—hence the name "wind-scorpion." Most solifugids have a crepuscular or nocturnal pattern of activity, but a few, such as the North American *Hemerotrecha californica*, are day-active. Periods of inactivity are spent in burrows excavated in the soil, and it is here also that molting occurs.

Solifugids are voracious predators, particularly as juveniles and adult females. The diet consists of various insects: grasshoppers, beetles, flies, and, especially in the case of American species, termites. Lizards may also be taken, but it is the predilection for termite prey that is perhaps the most interesting from an ecological point of view, since termites are important primary decomposers in arid environments. The extent to which such predation may regulate termite populations deserves further attention. In this context it is appropriate to anticipate what is to come later by mentioning the fact that termites are also preyed upon by harvester ants and spadefoot toads. The exploitation of this common resource (termites) by arachnids, insects, and anurans suggests that some mechanism for resource allocation may operate in regions where these groups are sympatric, although such a mechanism, if it exists, remains to be identified and investigated.

The reproductive biology of solifugids show some differences from, and some similarities to, the scorpions. These arachnids are not viviparous but produce egg masses that are deposited in soil burrows. Active nymphs appear after a period of about one month on the soil surface; they feed on insects, and they may also show cannibalistic behavior. Little is known about the longevity of these arachnids, although, intuitively, because of their large size they could be expected to have long life-spans and slow postembryonic development times. This apparently is not the case. According to Cloudsley-Thompson (1961b), the North African *Galeodes arabs* (Fig. 2.3) lives for only a year, is oviparous, and produces a relatively large number of young. North American species of the genus *Eremobates* have a life-span of less than a year, while members of the genus *Othoes*, inhabiting the deserts of northwest Africa, may live for two years (Kaestner 1968). These characteristics are suggestive of an opportunistic strategy that allows solifugids to exploit the most favorable conditions and avoid periods of environmental stress. However, as we will see in Chapter 6, these arachnids have relatively efficient mechanisms for conserving body water and so are able to tolerate, to some extent at least, drought conditions.

Uropygids

This group of predatory arachnids (Fig. 2.4) occurs throughout the southern United States and northern Mexico and also in the Far East; it is absent from the deserts of Africa and Australia. Uropygids are nocturnal in habit and, like the solifugids, can move very rapidly over the ground surface when disturbed. They are active on the surface mainly during the rainy season; periods of inactivity are spent in leaf litter, under stones or bark, or in tunnels excavated in moist soil. Unlike the solifugids, uropygids produce repugnatorial substances that protect them from predators. The vinegaroon, *Mastigoproctus giganteus*, of the New World uses a mixture of acetic acid and caprylic acid to repel attackers, whereas *Thelyphonus caudatus* of southern Asia employs formic acid as a deterrent (Kaestner 1968).

Uropygids, more familiarly known as whip scorpions because of the flagellarlike extension of the terminal abdominal segment, feed on woodlice, cockroaches, centipedes, and amphibians under laboratory conditions; little is known of their dietary habits in the wild. In view of their sparse occurrence in the most arid of environments, they are probably of little ecological importance here.

Whip scorpions lay eggs that are retained in a sac, attached to the gonopore of the female, and on hatching the young climb onto the mother's back and remain there until the next molt. Virtually nothing is known about the duration of the life cycle in these arachnids, although postembryonic development is relatively rapid, according to Kaestner (1968).

Fig. 2.3. The solifugid *Galeodes granti* (= *arabs*); male specimen. (Photo courtesy of J. L. Cloudsley-Thompson.)

38 Desert Soil Fauna

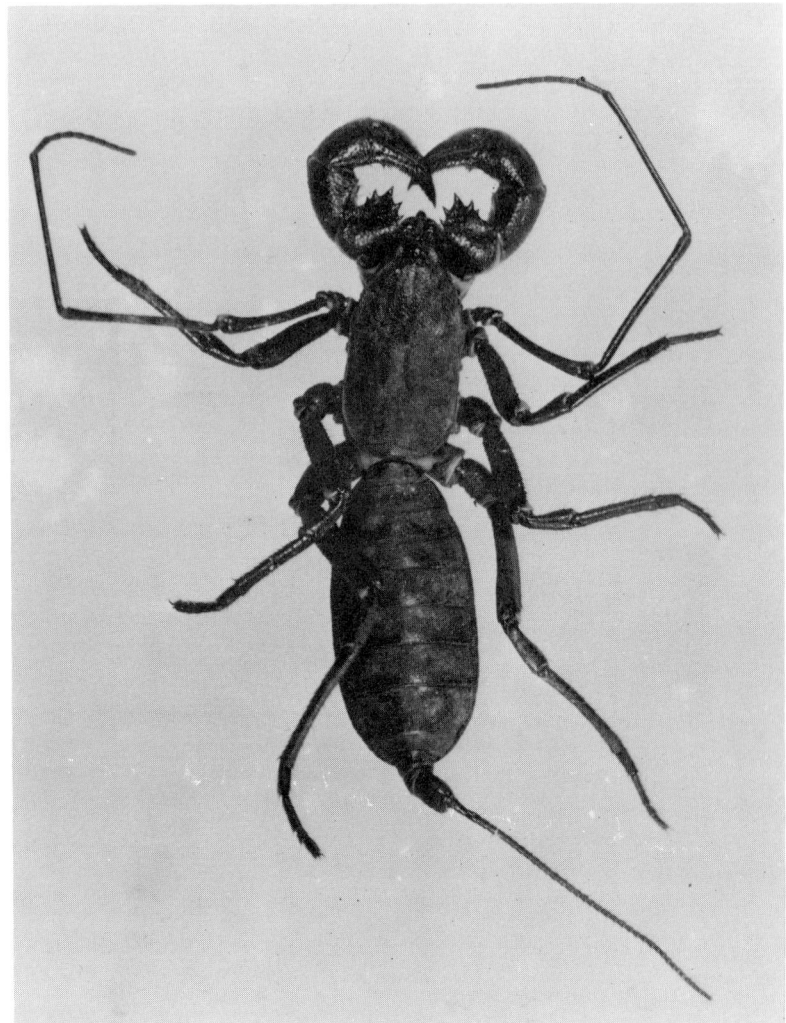

Fig. 2.4. The whip scorpion *Thelyphonus*. (Photo by E. A. Robins and reproduced courtesy of J. L. Cloudsley-Thompson.)

Spiders

Spiders are distributed widely in the deserts of the world, with a variety of families represented (Cloudsley-Thompson and Chadwick 1964). However, much of the information that is available about these desert dwellers is descriptive and, perhaps with the exception of the tarantulas and trapdoor spiders, is mainly concerned with species that cannot really be considered as members of the soil fauna at all. Riechert and her co-workers have reported some elegant studies on web-site selection by the funnel-web spider *Agelenopsis aperta* in the Chihuahuan desert grassland of New Mexico (Riechert, Reeder, and Allen 1973; Riechert 1974; Riechert and Tracy 1975; Riechert 1976). Unfortunately, the temptation to review these studies must be resisted since *A. aperta* lives almost entirely above ground and feeds for the most part on aerial insects such as dipteran Sarcophagidae and homopteran Cicadellidae (Riechert and Tracy 1975). These activities would seem to have little impact on soil-based biological systems.

Lycosa carolinensis, on the other hand, burrows into the soil in the uplands of the Sonoran desert. This species can withstand high summer, and widely varying seasonal temperatures (Moeur and Eriksen 1972), as we will see later (Chapter 7). Beyond this we can say very little. Nothing appears to be known, for example, about its feeding habits, impact on prey populations, vulnerability to mortality factors, reproductive tactics to combat mortality, and longevity in an unpredictable environment. *Agelenopsis aperta* has an annual life cycle, with few adults surviving the floods occasioned by summer rains (Riechert 1974). However, observations on this species cannot, with conviction, be extrapolated to *L. carolinensis* in view of the different ecologies of the two species.

In a survey of the arthropod fauna of a range of soil types in the Chihuahuan desert, Wallwork (in preparation) recovered spiders fairly consistently, but always in low numbers, from organic litter under saltbush, desert oak, juniper, and creosote bush. The identity of this fauna remains to be determined, and indeed there must be a question as to how representative these recoveries are of the entire soil spider fauna since the extraction methods employed (Tullgren funnels) select for small body size. Nevertheless, these observations indicate the presence of a subterranean spider fauna that perhaps is worthy of further attention.

So much for soil-based desert invertebrate carnivores. Attention can now be focused on those groups of invertebrates that qualify as primary consumers of the organic base that services the activities of desert soil communities.

MILLIPEDES

These myriapods are generally considered to be denizens of moist organic soils, particularly those of woodlands and forests. However, they have colonized arid and semiarid regions with varying degrees of success, and these forms are generally much larger in body size than their cool, moist temperate counterparts; there is a parallel here with the centipedes mentioned earlier. "Round-back" millipedes of the family Spirostreptidae occur in American deserts and also those of the Middle East and Africa. In Old World deserts this family is represented by the genus *Archistreptus*, while in the southwestern United States *Orthoporus ornatus* is the common desert millipede (Fig. 2.5), occurring in densities of about 1,300 ha^{-1} (Crawford 1976). Another group that inhabits arid and semiarid regions of Africa and Australia is the polydesmoid family Paradoxosomatidae. However, not all desert millipedes enjoy such a wide distribution. The Gomphiodesmidae, for example, is a family of round-back polydesmoids that is restricted to the African savannah (Lewis 1971a), a habitat that can be considered as marginally arid.

A common feature of desert millipedes is that they are active on the soil surface during the daytime but only for a restricted period of the year, which corresponds to the rainy season. The large spirostreptid *Orthoporus ornatus*, which is widely distributed in the Chihuahuan and Sonoran deserts, is dormant in the soil for eight to nine months of the year, retreating into hibernacula or the nests of the harvester ant *Novomessor cockerelli* (Fig. 2.6) during

Fig. 2.5. The desert millipede *Orthoporus ornatus* in the Living Desert State Park, Carlsbad, New Mexico.

Fig. 2.6. Nest of the harvester ant *Novomessor cockerelli*; a refugium for the desert millipede *Orthoporus ornatus*. (Photo by W. G. Whitford.)

the dry season (Wooten and Crawford 1974). The paradoxosomatids and gomphiodesmids of West Africa survive the dry season as diapausing larvae in spherical, subterranean molting chambers (Lewis 1971a, 1971b).

The features that allow millipedes to be active above ground during the day in hot deserts fall into two categories: behavioral and physiological. For the moment we are concerned with the former, and a consideration of these suggests that millipedes are, to some extent at least, preadapted to life in arid environments. The Temperate Zone millipede *Cylindroiulus punctatus*, for example, shows a marked seasonality in movements between the soil and habitats above the ground (Banerjee 1967), and such seasonality could provide a basis for natural selection. In hot deserts, species that restrict their periods of surface activity to times of the year when they would experience the least environmental stress would have a selective advantage. This is the rainy season, and, as we have seen, the species that have become established in hot deserts are those that are surface-active at this time. This would seem to be self-evident, but there is reason to believe that some millipedes are more successful than others in making this adjustment. An example of a relatively unsuccessful species is *Ommatoiulus moreletii*, an iuloid species that is indigenous in southwestern Europe. This millipede has been introduced into Australia where it has attained pest status in dry pastures. However, it suffers high mortality in grasslands during exceptionally hot and dry summers (Baker 1978) and clearly is at risk in this kind of environment. Its habit of aggregating in grass tussocks, rather than constructing hibernacula deeper in the soil, is probably the weakness in the defensive strategy of this species.

The more successful desert millipedes, of which *Orthoporus ornatus* is a good example, tend to remain close to their food sources during the hottest and driest parts of the day. These food sources (plant litter and bark—see below) are generally situated in shaded conditions where there is some measure of protection from climatic extremes. The West African paradoxosomatids *Habrodesmus duboscqui* and *Xanthodesmus physkon* feed, partly at least, on algae, and this habit may take them into shaded areas. In the Chihuahuan desert, algal films cover the soil surface on the margins of playa lakes and along the shaded edges of arroyos. Such sites remain moist longer than more exposed areas.

Daytime surface activity may also serve to lessen predation pressure since carnivorous arthropods, notably centipedes, scorpions, and their relatives, are nocturnal in hot deserts (see above). However, day-active predators are present in the form of birds and reptiles, which can pose a threat to millipedes active on the soil surface. On the other hand, it is a general feature of millipedes that they possess repugnatorial glands on at least some of the body segments. These glands produce a variety of noxious secretions that deter a would-be predator. Spirostreptids, for example, secrete benzoquinones, while polydesmoids produce benzaldehyde and hydrogen cyanide (Kaestner 1968). In this connection it is perhaps relevant to mention that larval paradoxosomatids exhibit swarming behavior on the soil surface (Toye 1967; Lewis 1971b). It has been suggested by Fryer (1957) that such swarms may serve to concentrate the odor of hydrogen cyanide and thereby ward off any predators.

Although carnivory has been reported in this group of arthropods (Hoffman and Payne 1969), the majority of millipedes feed on plant detritus, and in hot deserts they forage for food on the soil surface or in the aerial vegetation. Wooten and Crawford's (1975) study of the feeding habits of *Orthoporus ornatus* revealed a diet of dead leaves and bark of various desert shrubs such as creosote bush, mesquite, ocotillo, cholla, Russian thistle (*Salsola kali*), and Mormon tea (*Ephedra* spp.). The ability to utilize bark as a food source can be seen as an adaptation to surface feeding in hot deserts where shaded sites offer the most amenable microhabitats for activity.

According to Wooten and Crawford (1975), food selection by *O. ornatus* is very much a function of what is available in the habitat. This desert millipede is also very efficient in assimilating the food consumed (assimilation efficiency = 20 to 36 percent), and this is probably a reflection of the fact that it possesses a wide range of digestive enzymes (Nunez and Crawford 1976). In particular the presence of cellulases, pectinases, and xylanases in midgut secretions may enable *O. ornatus* to break down the cell walls and thereby utilize the cell contents of the plant material it ingests.

Orthoporus ornatus is long-lived; it has a life-span extending over several years (Nunez and Crawford 1977). In contrast the West African gomphiodesmid millipedes studied by Lewis (1971a) complete their life cycle in just over two years, and the adults live for only a few months. This is true of *Tymbodesmus falcatus*, which, although subterranean for the most part, occurs on the soil surface in the Nigerian savannah during the rainy season as late larval stadia and newly molted adults (particularly males). Copulation occurs on the soil surface, and eggs are laid in the soil during the wet season. The larvae that hatch from these eggs develop through seven stadia during the dry season, and stadium VII is surface active during the subsequent rainy season. This stadium reenters the soil and passes the next dry season in molting chambers. With the advent of the summer rains it molts to the adult, which emerges onto the soil surface and after copulation and egg laying dies before the end of this season. This life cycle is summarized in Fig. 2.7.

Another West African gomphiodesmid, *Sphenodesmus sheribongensis*, differs from *Tymbodesmus falcatus* in that only the adults are surface-active. This is a smaller species than *T. falcatus* and clearly spends more time below ground. However, the adults of *S. sheribongensis* are as short-lived as those of *T. falcatus*, since they appear at the beginning of the rainy season and die out by the end of it.

The Nigerian paradoxosomatid millipedes, *Habrodesmus duboscqui, Xanthodesmus physkon,* and *Xanthodesmus* sp., studied by Lewis (1971b) are surface dwellers for the most part, burrowing in the soil only to molt, lay eggs, and overwinter. The life cycle takes one to two years in *H. duboscqui* and *Xanthodesmus* sp. and is completed in one year in *X. physkon*. The dry season is passed as diapausing larvae, protected in molting chambers in the soil, and the short-lived adults die after egg laying and before the end of the rains. The eggs are laid in a batch in the soil, often consisting of

44   Desert Soil Fauna

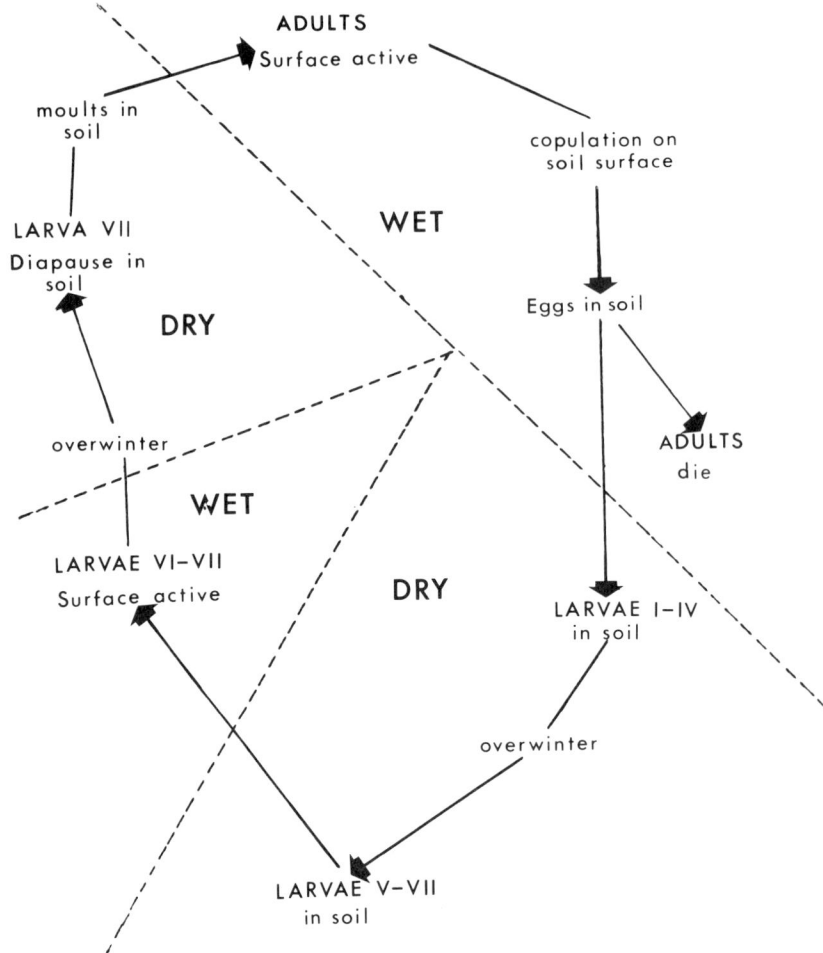

Fig. 2.7. The life cycle of *Tymbodesmus falcatus*. (From Lewis 1971a.)

over 1,000. On hatching the larvae appear on the surface and show swarming behavior. More often than not, these swarms comprise a single stadium, and Toye (1967) suggested that swarming may be a device to reduce water loss. Lewis (1971b), on the other hand, advances the opinion that swarms of early stadia that, in the species studied, are pigmented, may provide a form of camouflage to protect against predation. Earlier in this chapter attention was drawn to the possible deterrent effect of such swarming behavior. Clearly, the last word has not yet been written on this story.

WOODLICE

These isopod crustaceans are among the most unlikely candidates for life in arid soils since they belong to a group of arthropods that has achieved its greatest success in marine and freshwater environments. The woodlice are pioneers on land where they are generally restricted to the cool, moist microsites of bark crevices and the leaf litter on which they feed in temperate deciduous forests. This restriction is imposed because the cuticle covering the body surface does not usually possess a waterproofing layer of lipid of the kind present in scorpions and insects. The absence of such a layer can be interpreted as a sign that woodlice have not become fully terrestrial—that they have not yet evolved a key adaptation that would allow them to colonize dry habitats. Such a waterproofing device would, of course, drastically reduce the amount of water lost across the body surface by transpiration. On the other hand, it has been argued (Cloudsley-Thompson 1969) that a permeable cuticle has positive advantages for woodlice in that it allows evaporative cooling to occur during periods of exposure to high temperatures. It also allows excess water, taken into the body during times of flooding, to be eliminated. However, it must be pointed out that those woodlice species that have become established in arid environments do not appear to employ evaporative cooling to counteract high temperatures (Warburg 1968; see also Chapter 6). Again, the movement of water across the body surface is a two-way process, and if woodlice had been able to develop an epicuticular water barrier, this would prevent uncontrolled uptake during periods of immersion. There would then be no need for a pathway of elimination.

Placing these arguments aside for the moment—we will return to them in Chapter 6—the fact remains that the isopod cuticle has a greater degree of permeability than that of many other desert arthropods. Consequently, the biology of these crustaceans in hot deserts is of considerable interest.

The woodlice species that have been most successful in establishing themselves in hot deserts—and on which most attention has been focused to date—belong to two families: the Porcellionidae and Armadillidae. The porcellionids include *Porcellio olivieri* and *Hemilepistus reaumuri* of the North African and Middle East deserts (Fig. 2.8), while the armadillids are represented by such species as the Middle Eastern *Armadillo officinalis* and *A. albomarginatus* and the rare North American *Venezillo arizonicus*. Mention should also be made at this point of *Periscyphis granai*,

Fig. 2.8. The desert woodlouse *Hemilepistus reaumuri* commencing a burrow excavation in the Negev desert. (Photo courtesy of C. S. Crawford and reproduced by kind permission of Springer Verlag.)

which occurs in the western highlands of the Saudi Arabian desert in densities ranging from 100 to more than 600 m^{-2} (Kheirallah 1979).

The relative importance of these woodlice in desert ecosystems varies considerably. *Venezillo arizonicus* clearly can have only a marginal impact on the functioning of such communities in view of its rarity. *Hemilepistus reaumuri*, on the other hand, may play a key role in the herbivore/detritivore food chains in the Negev desert of Israel. Data collected by Yossi Steinberger (1980: personal communication) show that the biomass of this isopod is considerably higher than that of other herbivores in this desert (Table 2.2). *H. reaumuri* and *V. arizonicus* may well represent two extremes of commonness and rarity, and within the limits of these extremes are probably located the other desert woodlice: *Armadillo officinalis, A. albomarginatus, Porcellio olivieri,* and *Periscyphis granai.*

Hemilepistus reaumuri is unusual in being crepuscular; most desert wood lice are nocturnal, living under stones or rock crevices during the day. *H. reaumuri* (Fig. 2.8) inhabits loess soils in the Saharan and Negev deserts where it constructs vertical burrows

TABLE 2.2.
Biomass of Common Herbivores on Loess Plain of the Negev Desert

	Biomass (g 1,000 m^{-2})
Isopods (*Hemilepistus reaumuri*)	1,000
Porcupines (*Hystrix indica*)	150
Jerboa (*Jaculus jaculus*)	70
Hares (*Lepus europeus*)	2.5

Source: Steinberger unpubl.

(Fig. 2.9) into which it can retreat. As a result of the work of Shachak and his colleagues (Shachak, Chapman, and Steinberger 1976; Shachak, Steinberger, and Orr 1979; M. Brown, unpubl.), a good deal of information has been accumulated on the biology of this species.

H. reaumuri, one of 10 species belonging to the subgenus *Hemilepistus*, is a social isopod. It lives in family units of between 60 and 150 individuals, and the integrity of each unit is maintained by pheromonal "badges" that allow intruders to be identified and ejected (Linsemair 1972; Shachak, Steinberger, and Orr 1979). The focal point of family activity is the burrow system, which is vital for survival. The Negev desert is a region of hot summers, cool winters, and a low and consistent, regular rainfall (Shachak 1980). Survival in this environment depends on the ability of the wood louse to find conditions in which the moisture content of the air does not fall below 6 percent; during a drought period such conditions may only occur at depths of between 40 and 50 cm below the soil surface. It is imperative, therefore, for the burrow system to be excavated to such depths if the family is not to be wiped out. An excavation on this scale requires the participation of all members of the family, juveniles as well as parents, and the maintenance of the family unit spells the difference between success and failure. To see more clearly how this strategy operates, it is necessary to review the main features of the life history of *H. reaumuri*.

This isopod has an annual life cycle. Pair formation occurs on the soil surface from February to the end of March or the beginning of April in the Negev. During May the young hatch and emerge from the brood pouch. By this time the female parent has constructed a shallow burrow (to a depth of about 10 cm), and the young remain underground for about two weeks. After this time they emerge onto the soil surface in the morning and evening to forage, like their parents, for green and dead plant material and algal films. By the end of July the young are almost full grown, and

48 Desert Soil Fauna

Fig. 2.9. The burrow system of *Hemilepistus reaumuri* in the Negev desert. (Photo courtesy of C. S. Crawford and reproduced by kind permission of Springer Verlag.)

they may continue to be surface-active until November. At this time, and during the following winter months, the parents die, the new generation remains below ground, and the entrance to the burrow is sealed off. The members of this generation emerge from the burrow at the time of the next (February) rains, disperse, form pairs, and commence the life cycle over again (Fig. 2.10).

The initial burrowing activity of the pregnant female during the period from February to April/May produces a shallow vertical shaft. While the female is so engaged, the male will often mount guard over the developing burrow and assist in the removal of feces from the burrow. These feces are deposited in a characteristic

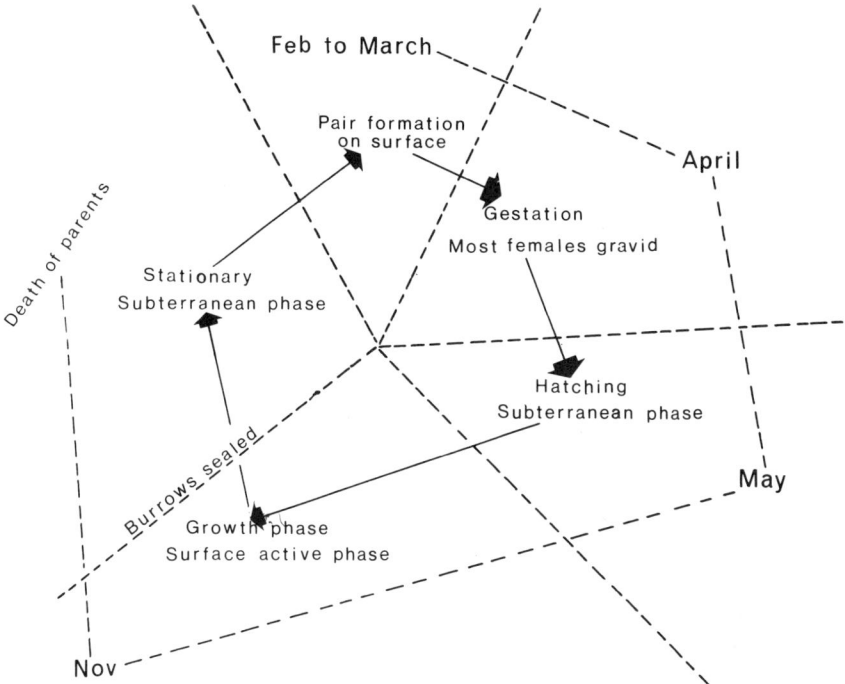

Fig. 2.10. Life cycle of *Hemilepistus reaumuri* in the Negev desert. (After Shachak 1980.)

saucer-shaped mound around the burrow entrance (Fig. 2.9). From June onward the developing juveniles participate in the excavation of the burrow that extends not only downward but also laterally as tunnels and rooms. This progression is depicted in Fig. 2.11.

Hemilepistus reaumuri is, as we have noted, an unusual desert isopod; it produces an elaborate burrow system in the soil; like the related *H. afghanicus* from Afghanistan it is social; it is not strictly nocturnal and, in this regard, shows a positive photoreaction except at high temperatures (Warburg 1968). Other desert woodlice do not construct soil burrows, and they show an avoidance reaction to light, for the most part. It is now time to pay some attention to these.

During their times of inactivity many woodlice exhibit clustering behavior, which is evidently a device for restricting water loss from the platelike pleopods carried on the undersurface of the abdomen. The armadillids *Aramadillo officinalis* and *A. albomarginatus*, have gone a stage further by developing the ability to conglobate—that is, to be able to roll the body into a ball, armadillo

50 Desert Soil Fauna

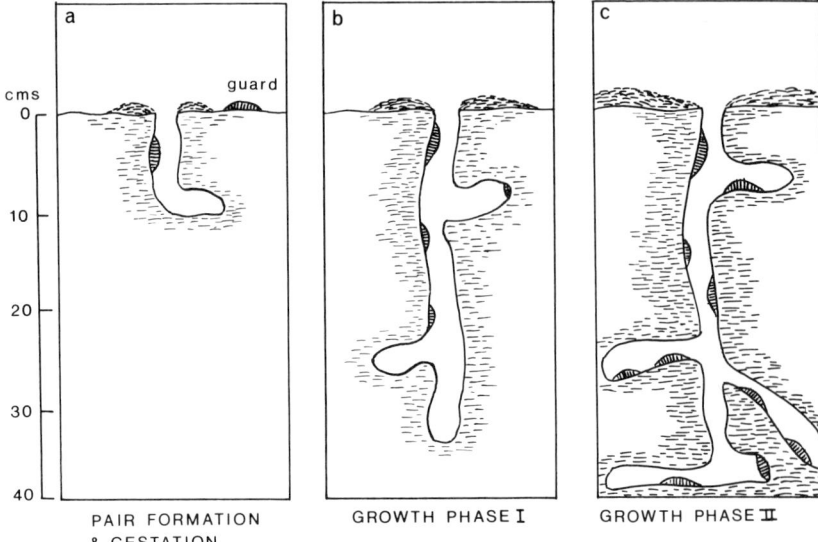

Fig. 2.11. Development of the burrow system of *Hemilepistus reaumuri* through its annual life cycle. (After Shachak 1980.)

fashion. This ability is also possessed by *Armadillidium vulgare*, a species that is not a true desert dweller but that occurs frequently on the margins of deserts in semiarid grassland.

Hemilepistus reaumuri and *H. afghanicus* are surface active for 9 or 10 months of the year, but such prolonged periods of activity are probably unusual among desert woodlice. *Armadillidium vulgare*, for example, becomes inactive in deep soil crevices during the dry season in semiarid grassland (Paris 1963), and this may be a common pattern among desert dwellers that do not construct burrow systems. Even so, the *A. vulgare* populations studied by Paris (1963) emerged onto the soil surface at night during the dry season to feed, which suggests a degree of flexibility in their behavior pattern.

A good deal still remains to be learned about the life cycles of desert isopods. What little information there is suggests a pattern that departs radically from that shown by woodlice of cool, moist temperate regions. Thus, the nondesertic (although xerophilic) *A. vulgare* takes four to five years to complete its life cycle in a California grassland (Paris and Pitelka 1962), whereas, as we have just seen, *H. reaumuri* has a life-span of no more than a year (Schneider 1975; Shachak, Chapman, and Steinberger 1976; Crawford 1979b). In the same vein, *Armadillo officinalis* probably lives for no more than two to three years (El-Kifl et al. 1970) and, like *H. reaumuri*, breeds only once after attaining maturity. In contrast *Porcellio olivieri* in coastal Mediterranean deserts breeds early enough in the year (February to April) to allow rapid postembryonic development to provide sexually mature adults by June, and a second generation appears in the period between June and September (Kheirallah and Awadallah, in press). This same tactic is evidently adopted by *Porcellio laevis*, *Metaponorthus pruinosus*, and *Leptotrichus naupliensis*, which have life-spans of less than one year (El-Kifl et al. 1970). A more detailed discussion of the reproductive tactics adopted by desert woodlice is given in Chapter 9, but perhaps it is relevant at this point to make the distinction between the semelparous *H. reaumuri* and *A. officinalis*, which live for a year or longer, and the shorter-lived *P. laevis*, *M. pruinosus*, and *L. naupliensis*, which may be iteroparous.

SNAILS

Like the woodlice, snails belong to a phylum that has been much more successful in aquatic habitats than on land. However, pulmonate mollusks are widely distributed in terrestrial environments, particularly where moisture and calcium are available in some quantity. These commodities are in short supply in hot deserts, and so it is intriguing to find not only that a few species of snail have colonized these arid habitats but also that they tolerate, rather than avoid, the hot, dry conditions. The mechanisms whereby this tolerance is achieved are considered in more detail in Chapters 6 and 7. For the moment it is sufficient to point out that the spirally coiled shell of the snail provides not only a measure of protection from predators but also a means of reducing water loss from the moist body surface. In addition it may create a buffer zone that ameliorates the effects of high ambient temperatures. The aperture of this shell can be blocked by the mantle collar or sealed off by the secretion of an epiphragm during periods of environmental stress; in some cases the epiphragm can seal the shell to the substratum.

Desert snails are relatively small in size, a feature that enables these animals to retreat into moist microenvironments and also to seal efficiently against irregular rock surfaces (Mazek-Fialla 1934) and aerial vegetation. The Middle Eastern *Trochoidea seetzeni*, for example, estivates during the summer months and attaches to shrubs by its calcareous epiphragm (Yom-Tov 1972). However, *Sphincterochila zonata* (= *biossieri*), which is perhaps the best adapted of desert snails, has developed neither of these behavioral traits to any significant extent (Machin 1967; Schmidt-Nielsen, Taylor, and Shkolnik 1972). It has, however, structural and physiological features to compensate for this (see Chapters 6 and 7). Much the same is true of the North American *Rabdotus schiedeanus* (Riddle 1975).

The genus *Sphincterochila* has a Mediterranean distribution and is represented by five species in the deserts of Israel and Sinai (Bar 1975), namely, *S. cariosa*, *S. fimbriata*, *S. aharonii*, *S. zonata*, and *S. prophetarum*. Habitat partitioning between these species is considered in Chapter 8; for the moment our attention is focused on *S. zonata* and, to a lesser extent, *S. prophetarum*, which have been more extensively studied than the others. *S. zonata* is native in the deserts of the Negev, Jordan Valley, and northern Arabia (Fig. 2.12) where it is sympatric with *Trochoidea seetzeni* on such substrata as firmly packed loess, limestone rock and pebbles, and fragmented flints. Neither species occurs on sandy soil. However, *S. zonata* and *T. seetzeni* have different ecologies, which allow them to coexist (Schmidt-Nielsen, Taylor, and Shkolnik 1972).

Desert snails spend long periods of time in a dormant state, which is economical in terms of energy costs and water utilization. Periods of inactivity are spent under bushes or large rocks where,

Fig. 2.12. The desert snail *Sphincterochila* in the Negev desert. (Photo courtesy of C. S. Crawford and reproduced by kind permission of Springer Verlag.)

in the case of *S. zonata* at least, the snail may bury itself in the mounds of loess that accumulate around the base of desert shrubs. These mounds apparently are less susceptible to waterlogging during the rains than is the surrounding soil, and yet they retain their moisture longer (Shachak, Orr, and Steinberger 1975). *S. zonata*, *S. prophetarum*, and *T. seetzeni* become active during the rainy season, and it is at this time that feeding activity is most pronounced. In the case of *S. zonata* and *S. prophetarum* this period lasts for 20 to 25 days in the year (Shachak, Chapman, and Orr 1976; Yossi Steinberger 1980: personal communication). Snails are mainly herbivorous animals, feeding on living green plant material, which they are able to digest through the presence of cellulases in their gut enzyme compliments. *Trochoidea seetzeni* conforms to this pattern by feeding on the shrubs within which it finds refuge. *S. zonata*, on the other hand, is mainly a mud feeder, and its feeding activity is more restricted to times when soil is sufficiently moist to be acceptable as a food source. These two examples illustrate the subtle differences in resource partitioning that may exist between two species of desert snail; there may well be others, which will provide fruitful areas for future research.

The prolonged periods of inactivity characteristic of desert pulmonate mollusks, with their consequent economy of maintenance costs, are associated with long life-spans. *S. zonata*, for example, can survive for at least three years, the sympatric *Eremina desertorum* for more than four years, and the North American *Xeranianta veatchii* for six years (Machin 1967).

Data on the reproductive biology of desert mollusks are hard to find, and much of what is available relates to species inhabiting the deserts of the Middle East. *Spincterochila zonata* mates soon after the onset of the summer rains (December) in the Negev (Shachak, Orr, and Steinberger 1975). Mating occurs during the daytime, and eggs are laid after a gestation period of a month in a cavity excavated in the soil. Hatching occurs within the egg burrow, and the young snails emerge on to the soil surface during the next rainy season. Here they are subject to predation by thrushes, which may take up to 27 percent of the population. Density-dependent control of reproduction in *Trochoidea seetzeni* is discussed in more detail in Chapter 9. There, for the moment, we leave it.

INSECTS

The insects are superbly adapted to life on land. They are highly mobile, both on the ground and in the air; they have an epicuticular lipid layer that restricts water loss across the body surface; their excretory organs (Malpighian tubules) are designed to minimize water loss during the elimination of nitrogenous wastes, and their tracheal method of respiration, with efficient spiracular closing mechanisms, controls this avenue of water loss. These are just some of the features of insects that can be utilized by would-be colonizers of arid environments.

It is often stated that insects are the dominant group of terrestrial arthropods both in terms of sheer numbers and species diversity. But this needs some qualification when we look at soil communities. To be sure there are many insect groups occurring in this system, but there is some question as to the extent to which they dominate or regulate ecosystem processes, notably those concerned with decomposition. This is particularly pertinent to desert soil systems, as will become apparent later. For the moment, however, attention is focused on those groups of insects that form part of the primary consumer fauna, based in the soil and active on its surface in hot deserts. These insects belong mainly to four Orders: Dictyoptera, Isoptera, Coleoptera, and Hymenoptera. Although these are considered here to be surface-active herbivores, granivores, or detritivores, it is recognized that this is an overgeneralization. Some termites are completely subterranean in habit, some are cannibalistic; some harvester ants that feed on seeds can be predaceous on occasion. This topic of "diet switching" is discussed in Chapter 8; for present purposes it is convenient to treat these polyphagous groups as entities.

Dictyoptera

Cockroaches are successful colonizers of arid regions in Australia and northern Egypt (Mackerras 1970; Ayyad and Ghabbour 1977; Ghabbour, Mikhail, and Rizk 1977), but little is known of their biology. This is not the case, however, with the desert cockroach of North America, *Arenivaga*. This has been studied intensively and provides a fascinating example of adaptation to the extreme environment in which it lives.

Arenivaga (probably *A. investigata*) (Fig. 2.13) lives in the sandy deserts of the southwestern United States and northern Mexico where it burrows in loose sand and feeds on detritus. It is well equipped for this mode of life, having a flattened, discoidal body and legs furnished with strong spines. The adult males are

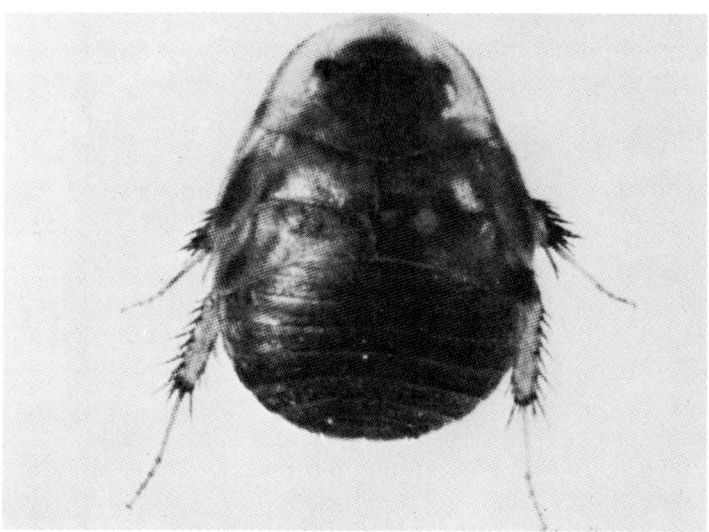

Fig. 2.13. The desert cockroach *Arenivaga investigata* from the Mojave of southern California. (Photo by G. Ott.)

winged, but the females and, of course, the nymphs are wingless. This cockroach also occurs in desert grasslands where it can be encountered in the soil burrows of vertebrates (Crawford 1979b). The key to its success in hot deserts rests on its ability to do two things: to exploit, to the maximum, the sources of environmental moisture that are available and to conserve this moisture in the body once it has been obtained. The ways in which *Arenivaga* achieves these are dealt with in Chapter 6.

In the Mediterranean desert of Egypt the sand roach *Heterogamia syriaca* is a conspicuous member of the soil mesofauna (Ghabbour, Mikhail, and Rizk 1977). In this region *Heterogamia* apparently has two reproductive seasons, one in January/February (the rainy season) and the other in August when soil moisture levels are, evidently, relatively high. Ghabbour, Mikhail, and Rizk (1977) reported that population densities of *H. syriaca* were more closely correlated with soil moisture at a depth of 50 cm than at 20 cm. This led these authors to conclude that the roach burrows deeply into the soil during the daytime but moves upward to the surface at sunset to feed, mate, and take advantage of the water available as dew formation occurs. A similar pattern of activity is apparently shown by *Arenivaga investigata* (Edney, Haynes, and Gibo 1974).

Isoptera

Termites are frequent in arid environments, although they have radiated less extensively here than in tropical forests, woodlands, and savannahs (Lee and Wood 1971; Brian 1978). Nevertheless, they are important consumers of net primary production in hot deserts (Crawford 1979b). They are social insects that construct nests or mounds, often elaborate, above or below the surface of the soil. The size of colonies varies widely from several hundred or a few thousand individuals to several million. Perhaps a typical estimate of density in hot deserts is that provided by Schaefer and Whitford (1981) for the Chihuahuan desert species *Gnathamitermes tubiformans*. These authors estimated an average of 80 colonies per hectare, with an average colony density of 10,000 individuals. It was also calculated that 48,000 (6 percent) of the 800,000 termites per hectare were actively engaged in foraging at any one time.

The termite species that live in hot deserts tend to construct subterranean nests. This is true of the most abundant species in the Sahara, *Psammotermes hybostoma* and *Anacanthotermes ochraceus*, for example. In a semiarid mallee of southern Australia, Lee and Wood (1971) encountered nine species of termite, all of which were subterranean. In western Australia, on the other hand, mound builders and subterranean species occur sympatrically. Mound builders include the harvester termites, such as the Australian *Drepanotermes perniger*, which forage on the surface, usually at night. Some subterranean termites, such as *Psammotermes hybostoma*, construct concentrated nest systems with extensive galleries. Many of these species forage entirely within the soil, although *P. hybostoma* is polyphagous and feeds on windblown debris, dung, and a variety of plant material. Other species form small colonies under rocks or surface accumulations of organic debris (Fig. 2.14). Some of these are subterranean foragers, but others, such as the Chihuahuan *Amitermes* and *Gnathamitermes* species, emerge on to the soil surface to forage on dead annuals when saturation deficits and surface temperatures are low (Whitford 1978a). Here they are heavily preyed upon by harvester ants belonging to the genera *Pogonomyrmex*, *Novomessor*, and *Pheidole*.

There is a considerable amount of information in the literature concerning the feeding biology of termites (see, for example, the recent detailed review by LaFage and Nutting 1978), but again much of this does not relate directly to species living in hot deserts. What is available suggests that desert termites select cellulose-rich substrates, and these are provided for them by dead wood, the dung

Fig. 2.14. Accumulations of organic debris (ironwood and Palo Verde) being worked by termites in the Sonoran desert near Casa Grande, Arizona.

of desert cattle, and the woody frass produced by the feeding activities of wood-boring beetle larvae. Nutting, Haverty, and LaFage (1975) consider termites to be the most important consumers of dead wood in hot deserts, and dry-wood termites nest within their food supply until it is decomposed to a point where it becomes unsuitable as food (LaFage and Nutting 1978). Certainly the dry-wood termite *Pterotermes occidentalis* would seem to utilize the sawdustlike frass of wood-boring beetle larvae since its galleries are often driven through the tunnels of these insects. On the other hand, although attention has been drawn (see above) to the fact that *Amitermes* species will forage on dead annuals, 32 out of 46 termite species that feed on cattle dung in Australia belong to this genus (Ferrar and Watson 1970). Similarly, *Gnathamitermes perplexus* commonly feeds on cattle dung in the Sonoran desert (LaFage and Nutting 1978). Studies on the related *G. tubiformans* and on *Amitermes wheeleri* in the Chihuahuan desert by Ettershank, Ettershank, and Whitford (1980) indicate that these termites can locate prospective food items much more easily if these are on the surface of the soil rather than buried in it. Food items, such as dung pats, lying on the surface create temperature anomalies in the soil profile, and the detection of these irregularities by the termites may be a way of locating their food.

In view of their large densities and their ability to feed on a wide variety of organic substrates, desert termites must have an important role in ecosystem functioning. This topic has been addressed by Schaefer and Whitford (1981) with particular emphasis on the activities of these insects in relation to nutrient cycling. They point out that subterranean termites transport large quantities of dead plant material into their soil galleries, often to depths beyond the reach of plant rooting systems. Further, the nutrients assimilated into termite tissue tend to be recycled here by trophylaxis and cannibalism. The release of these nutrients from this ecological sink is mainly through the predations, on termite workers and alates, by ants, amphibians, lizards, and birds. In addition, such desert termites as *Gnathamitermes tubiformans* can fix atmospheric nitrogen and use this in the production of gallery "carton"; this source of nitrogen, deployed in surface galleries, will be available to plants.

The removal of surface litter into underground sites by termites is a key industry in desert grasslands since it has been calculated that these insects remove about 55 percent of this litter in such areas in west Texas (Bodine and Ueckert 1975). A similar level of activity has been reported for subterranean termites in creosote bush and mesquite-yucca grasslands in the Chihuahuan desert (Johnson and Whitford 1975). In these sites about 70 percent of the organic input to the soil is contributed by leaves of the creosote bush. This material does not rank high on the list of preferred food for termites, and, indeed, in wet years the diet consists mainly of litter produced by annual grasses and forbs. However, in dry years when this material is in short supply, desert termites turn their attention to creosote bush litter (Walter Whitford 1980: personal communication). This observation suggests that the activity of termites as primary consumers of dead plant material is somewhat variable in arid localities. Indeed, Wood and Sands (1978) have compiled data that indicate decreases in the number of termite species, their overall abundance, and the proportion of detritivores to other feeding types along an environmental gradient from humid rain forest to semiarid Sahel savannah in Africa.

Bouillon (1970) has shown that termites, in general, correlate their egg-laying, swarming, nest-building, and foraging activities with seasonal climatic events, although this generalization may not apply in hot deserts. Nutting and Haverty (1976), reporting on desert species, noted that the nuptial flights of the winged reproductive caste were not timed to coincide with seasonal rainfall, which suggests a measure of independence of physical environmental factors. However, these authors agree with others (Watson and Gay 1970; Johnson and Whitford 1975, for example) that temporal and spatial distributions of termites are limited by seasonal and annual fluctuations in climate. Clearly there is a need for much more information on the phenology of termites, in general, and desert-dwelling species, in particular.

Termite colonies vary considerably in their longevity; some colonies achieve maximum size in nine months, whereas others take 25 to 30 years to achieve this (Lee and Wood 1971) and may have a longevity of 50 years or more. Such observations relate to mound-building species and may not be directly applicable to the smaller, subterranean colonies prevalent in hot deserts. Crawford (1979b) places these in what he calls the "intermediate life history" grouping and notes that their longevity varies considerably with the caste; workers are short-lived, whereas reproductives may live for many years. But he gives no further details.

Coleoptera

Surface-active beetles are among the most conspicuous members of the fauna associated with soil in many temperate and tropical regions, and hot deserts are no exception. The most frequent members of this fauna in cool, moist temperate grasslands and forests belong to the families Carabidae and Staphylinidae, both of which include carnivorous and detritivorous species. In hot deserts, however, carabids and staphylinids are largely replaced by members of the Tenebrionidae and, around animal carcasses and dung, the Scarabaeidae. Other families of beetles that are associated with dung and carrion in hot deserts include the Trogidae, Geotrupidae, Dermestidae, and Silphidae, while others, such as the Lathridiidae, are fungal-feeding inhabitants of leaf litter. Virtually nothing is known about the ecology of these desert insects. The

60 Desert Soil Fauna

Tenebrionidae, which are widely distributed within and between deserts, have received more attention than most, and what follows concerns them.

Desert tenebrionids (Fig. 2.15), as a whole, show a wide range of activity patterns, both circadian and seasonal. Edney (1971), for example, was able to distinguish, among the tenebrionids of the Namib desert, species that were day-active from those that were nocturnal or crepuscular. Superimposed on this circadian pattern were seasonal differences such that day-active and crepuscular species were primarily abroad during the summer, whereas nocturnal species confined their periods of activity to the winter (see Chapter 8 for further discussion). Periods of inactivity are spent in a dormant condition in soil burrows.

These differences in activity patterns are very much related to differing abilities to tolerate desert conditions, in particular the ability to control water loss. The water relations of desert tenebrionids have been studied in some detail and they are discussed in Chapter 6.

Desert tenebrionids are detritus feeders, particularly as adults, which feed on the soil surface, taking particles of organic debris of plant and animal origin. *Cryptoglossa* sp. inhabiting barchan dunes in the Mojave desert, for example, feeds mainly on dead bees

Fig. 2.15. A Negev desert tenebrionid, with *Hemilepistus* in the background. (Photo courtesy of C. S. Crawford.)

and the feces of the kangaroo rat *Dipodomys* (Martz 1980). These food items are blown about by the wind and collect on the lee side of the dunes where the foraging activities of *Cryptoglossa* are concentrated. In the Chihuahuan desert tenebrionids congregate under logs, which probably represent an important food source, as well as a sheltered environment, and they move, ponderously, to other refuges when disturbed. Larvae of the North American *Eleodes obscurus* have a varied diet; they scavenge on dead plant material and animal remains, but they will also feed on roots and seeds (Wise 1979).

As Crawford (1979b) has noted, very little is known about the life histories of desert Coleoptera in general, and Tenebrionidae in particular. Many of the latter evidently have a long period of postembryonic development, (James McBrayer 1979: personal communication), but there are some species that are short-lived, (Clayton Gist 1980: personal communication) at least as adults. Again, many Coleoptera are only temporary members of the soil fauna in hot deserts, using this environment to pass through larval and pupal stages, while the adults are aerial (see Chapter 4). One interesting example of this is provided by a chrysomelid beetle that lives in the White Sands National Monument of New Mexico. The larvae of this beetle inhabit the leaf litter accumulations under saltbush and construct a case around them composed of sand grains. This is very reminiscent of caddis larvae, and although it is very unusual among beetles, it is a habit adopted by members of the genus *Cryptocephalus*. Clearly, it is an adaptation to life in a very dry environment. There must be many more as yet undiscovered adaptations among larval Coleoptera that live in hot deserts.

Hymenoptera

The most successful group of desert hymenopterans is unquestionably the ants. These social insects live in subterranean nests, and they forage on the surface either singly or in groups. The extensive studies carried out by Whitford and his co-workers (Schumacher and Whitford 1974, 1976; Whitford and Ettershank 1975; Whitford 1976a) provide some idea of ant densities in hot deserts. The number of colonies per hectare can vary from one (in the case of the predatory *Myrmecocystus mexicanus*) through hundreds (harvester ants, such as *Pogonomyrmex, Novomessor*, and *Pheidole* species) to several thousand, in the case of small-sized omnivores such as *Iridomyrmex pruinosum* and *Conomyrma* spp.

62 Desert Soil Fauna

Total densities of ants within a particular colony are difficult to determine; instead, estimates are more often expressed as the number of active foragers per colony. More often than not, these estimates fall within the range of 1,000 to 4,000, and if these are multiplied by the number of colonies per hectare, ant densities on this area basis can vary from 1,000 to nearly 3 million, depending on the feeding "guild" considered (see below). The diversity of desert ant faunas is also considerable. Chew (1977), for example, recorded at least 26 species from a bajada site in southern Arizona, and this is probably typical of many hot desert sites.

As with the termites, the nest of the desert ant (Fig. 2.16) provides an admirable physical refuge into which these insects can withdraw when surface temperatures become intolerable; it also provides a site for the storage of food collected on the surface. Ants are able to avoid, rather than tolerate, the physical extremes of the desert environment. This is illustrated by their patterns of surface activity.

Ants show a great deal of flexibility in their patterns of surface activity. Daytime activity is the rule rather than the exception during the cooler months of the year, but many species switch to nocturnal or crepuscular patterns during the warmer months. This is particularly true of those species belonging to the

Fig. 2.16. Saucer-shaped mound surrounding the entrance to the nest of the harvester ant *Pogonomyrmex*.

genera *Pogonomyrmex, Pheidole, Solenopsis,* and *Veromessor,* which forage in groups (Brown, Reichman, and Davidson 1979). In contrast species of *Pogonomyrmex* that forage individually, such as *P. desertorum* (Whitford 1978b), are essentially day-active throughout the year, and Davidson (1977a) has suggested that this is a reflection of their need for visual orientation. Some ants are strictly nocturnal, and these include *Myrmecocystus mexicanus* and *M. navajo,* which have been variously labeled as honeydew-exudate feeders or largely predatory (Whitford 1978c). Nocturnal foraging is also practiced by the omnivore *Formica perpilosa,* and, here, nighttime activity may be a device to tap the pool of insect protein that is available in hot deserts at this period during the 24-hour cycle, but it may also be a way of avoiding thermal stress and desiccation during the day (Whitford 1978c). *Pogonomyrmex rugosus* and *Novomessor cockerelli* have a bimodal activity pattern, taking seeds during the day and insects at night (Whitford and Ettershank 1975; Whitford 1978c). Clearly desert ants display a variety of activity patterns, and the reader is directed to the works of Schumacher and Whitford (1974), Whitford and Ettershank (1975), and Whitford (1978c) for further documentation.

The diversity of feeding habit shown by ants of hot deserts is a good reflection of their adaptive radiation in these regions. Since foraging activity on the surface is the means by which food is acquired, this topic is closely linked with the previous one. For the moment it is sufficient to introduce the various types of feeding guilds represented among this group of hymenopterans. A discussion of the foraging strategies that permit a high degree of species packing in desert ant faunas is reserved for Chapter 8.

At an early stage in their evolution ants were, in all probability, active predators on other insects. Indeed, Stradling (1978) has suggested that most ant species are predaceous to some degree, and there is much in the literature to support this view (see, for example, P.J.M. Greenslade 1979; Greenslade and Greenslade, in press). Carnivory, as a way of life, is practiced by certain desert ponerines and a few species of the genera *Pheidole, Myrmecocystus, Dorymyrmex, Cataglyphis,* and *Neivamyrmex.* But evolution has produced lineages that have departed from this ancestral, carnivorous habit. The habit of feeding on wood has been adopted by the formicine *Camponotus.* Members of the genera *Myrmecocystus* and *Tapinoma* feed on plant juices and store these in the bodies of specialized "repletes." The seed-eating (gramnivorous or granivorous) habit has been taken up by a variety of myrmecines, notably members of the genera *Pogonomyrmex, Solenopsis, Messor, Veromessor, Novomessor, Pheidole, Holcomyrmex, Pheidologeton,* and

Meranoplus, although *Novomessor* and *Solenopsis* may more properly be regarded as omnivores (Whitford 1978c). These are the harvesters that often construct low, saucer-shaped mounds around the nest entrance (Fig. 2.16). Granivory is a key industry in desert ecosystems, and this feeding habit brings ants into competition with soil-based rodents (see Chapter 8). Finally, some ants are detritivores, and an example of this feeding guild in the deserts of the southwestern United States is *Trachymyrmex smithi neomexicanus* (Whitford 1978c).

One interesting aspect of the feeding biology of ants is the evolution of "vicariants," or similar guilds, in the deserts of North America, Africa, and Australia. For example, the granivore niche occupied by species of the genera *Pogonomyrmex, Pheidole, Solenopsis, Veromessor*, and *Novomessor* in North America embraces members of the genus *Messor* in Africa (Ghabbour, Mikhail, and Rizk 1977) and *Meranoplus* in Australia (Pisarski 1978). Certain species of the genera *Pheidole, Pogonomyrmex*, and *Myrmecocystus* are carnivorous in North America; this feeding niche is occupied by *Cataglyphis* and *Proformica* in Africa and *Myrmecia* species in Australia. Finally, honey-pot ants are present in the three regions, represented by species of *Myrmecocystus* in North America, *Proformica* in Africa, and *Melophorus* in Australia.

Independence of the external environment that sociality confers allows a great measure of stability for ant colonies as it does for termites. As a consequence the colonies of these social insects are long-lived. Chew (quoted in Brown, Reichman, and Davidson 1979) recorded colonies of *Novomessor* and *Myrmecocystus* persisting for more than 20 years. More specifically, this longevity can be attributed to the ability of ant colonies to store food, to be ectothermic, and to enter into a state of dormancy during periods of climatic stress and resource deficiency.

3

The Surface-Active Fauna: Vertebrates

Students of the soil fauna do not, as a general rule, consider vertebrates to be part of their brief. In hot deserts, however, a number of amphibians, reptiles, and mammals are based in the soil and have an impact on this ecosystem. They must, then, fall within the compass of this book. In introducing these elements of the soil fauna the practice adopted in the previous chapter is followed— namely to treat first the predatory groups, such as the marsupials and amphibians, and to deal last with the herbivorous lagomorphs. Conveniently linking these two extremes of diet are the reptiles, which on balance tend toward the predatory habit, and the rodents, which by similar tokens tend toward herbivory (or, more strictly speaking, granivory).

MARSUPIALS

The soil-based marsupials of the Australian deserts are essentially insectivorous, although some surface-active species may also consume small vertebrates such as lizards. These surface foragers belong to the family Dasyuridae, for the most part, and can be distinguished from the more truly subterranean marsupial moles (family Notoryctidae), which are considered in Chapter 4. Desert dasyurids, such as *Dasycercus cristicauda, Sminthopsis crassicaudata*, and *S. froggatti*, are nocturnal foragers, sheltering during the day in soil burrows. *D. cristicauda* the "mulgara," and *S. froggatti* inhabit arid, sandy areas where their diet includes both insects and lizards (Tyndale-Biscoe 1973). The fat-tailed marsupial mouse, *Sminthopsis crassicaudata*, on the other hand, avoids vertebrate prey and feeds entirely on insects and other arthropods; this species lives in more vegetated sites than those occupied by *D. cristicauda* and *S. froggatti*. In such sites arthropod prey will be more abundant than in open, sandy areas, but the returns on foraging will be proportionately less than those obtained by the larger-sized *D. cristicauda* and *S. froggatti*, which can take larger (vertebrate) prey more quickly. As a consequence *S. crassicaudata* forages intensively for its food and rarely enters into a state of torpidity, whereas the other two species assume a daily torpor and, thereby, curtail body water loss. The assumption of the torpid condition can be interpreted as a means of conserving energy, and this is evidently an important feature of the life-style of *D. cristicauda* and *S. froggatti*. This method of energy conservation apparently is not open to *S. crassicaudata*, but this species has the ability to store fat in its tail to a greater extent than these other species, and this may represent a temporary source of energy that could be drawn upon during periods of food stress.

All three of these desert marsupials avoid, to some extent, the stresses of the desert environment by seeking refuge in soil burrows and in this respect have a strategy in common with another marsupial, the insectivorous rabbit-eared bandicoot *Macrotis lagotis* (family Peramelidae). But the dasyurids have also evolved tolerance mechanisms that, in the case of *D. cristicauda* and *S. froggatti*, place emphasis on water gain from the environment (via their prey) and the restriction of water loss from the body by the assumption of the torpid state. Alternatively *S. crassicaudata* stores fat that will yield both energy and water when metabolized. These mechanisms are intimately associated with the ability to allow fluctuations in body temperature, and this is a topic that will be pursued in more detail in Chapter 7.

According to Tyndale-Biscoe (1973), *Sminthopsis froggatti* and *S. crassicaudata* are polyestrous; they breed when conditions are suitable and have a high reproductive potential. The litter size is about eight—which is larger on average than that of desert rodents (see Table 3.2). These patterns of continuous breeding are probably adaptations to life in an unpredictable environment, where aseasonality is the rule rather than the exception, and they contrast with those of temperate forest dasyurids, which are monoestrous and markedly seasonal.

AMPHIBIANS

It is perhaps surprising to record this group of vertebrates as members of the desert soil fauna in view of their reliance on a watery environment for the completion of their life cycle. Nevertheless, Warburg (1972) lists 38 species of anurans occurring in xeric or semixeric habitats throughout the world. He also notes that the majority of these belong to two families: the Bufonidae, which has a cosmopolitan distribution, and the Leptodactylidae, which are restricted to the Southern Hemisphere. In addition a few species belonging to the Pelobatidae (Northern Hemisphere) and Hylidae (cosmopolitan) have established themselves in arid environments. Much of what follows concerns these desert Anura, but perhaps we should not lose sight of the fact that urodeles such as the tiger salamander, *Ambystoma tigrinum*, also occur in deserts (Delson and Whitford 1973).

These desert anurans and urodeles are surface-active at night during the rainy season. At other times of the year they become dormant in soil burrows. The distribution of hylids is more closely linked with the occurrence of free water than is the case with the more terrestrial bufonids and pelobatids; leptodactylids occur in habitats ranging from aquatic to extremely arid. Here then we have a range of ecological types, not all of which can be considered to be well adapted to extreme desert conditions. Distribution patterns within any particular desert system suggest as much. In the deserts of the southwestern United States, for example, three species of the spadefoot toad genus *Scaphiopus* (*S. couchii*, *S. hammondii*, and *S. holbrookii*) occur, along with *Bufo cognatus*, in the most arid sites (Mark Dimmitt 1980: personal communication). These species are joined in less arid areas by other species of *Bufo*, such as *B. debilis, B. boreas,* and *B. punctatus* (Claussen 1969), as well as the hylid *Pternohyla* sp. (burrowing tree frog) and *Eleutherodactylus* sp. (barking tree frog) (Mark Dimmitt 1980: personal communication). The genus *Bufo* is also present on the northern fringes of the Sahara where it is commonly represented by the African desert toad, *Bufo regularis* (Fig. 3.1).

The spadefoot toads of the genus *Scaphiopus* (Fig. 3.2) have attracted the attention of desert biologists in recent years, and a good deal of information has been accumulated on their physiological ecology. A comprehensive list of references to this work, up to 1971, has been provided by Warburg (1972). Since that time, further contributions have been made by McClanahan (1972, 1975), Seymour (1973), Hillyard (1975), H. Brown (1976), Dimmitt and Ruibal

68   Desert Soil Fauna

Fig. 3.1. The African desert toad *Bufo regularis* in temporary pool. (Photo courtesy of J. L. Cloudsley-Thompson.)

(1980a, 1980b), and Shoemaker (1980), among others. This work is reviewed in Chapter 6.

Scaphiopus spp. are deep-burrowing forms, tunneling to depths of between 20 and 90 cm in winter (Dimmitt and Ruibal 1980a). In this respect they differ from the shallow-burrowing spadefoot toads of the Argentine desert (*Ceratophrys* spp.), which, however, are able to retain around themselves a parchmentlike cocoon as a form of protection during the drought. This phenomenon also occurs in Australian desert frogs (Shoemaker 1980).

Desert amphibians swarm in great numbers on the soil surface during the rainy season. The environmental clues that cause this emergence have been studied by Dimmitt and Ruibal (1980a). These workers found that in the case of *Scaphiopus couchii*, and probably also *S. multiplicatus*, emergence was stimulated by the vibrations produced by rain falling on the soil surface rather than by the presence of soil moisture—although this would also cause emergence. Surface activity after rains is most marked at night; during the day *Scaphiopus* retires to shallow soil burrows, 2 to 10 cm deep (Ruibal, Tevis, and Roig 1969). When active, the toads feed to build up energy reserves for use during the subsequent dormant

Fig. 3.2. The spadefoot toad *Scaphiopus couchii* from the Sonoran desert of North America. (Photo courtesy of M. Dimmitt.)

period and reproduce in the temporary pools that form after the rains. However, they are unable to exploit the temporary habitats of the playa lakes since the tadpole stage cannot tolerate the high salinity occurring here (Mark Dimmitt 1980: personal communication). In other sites, however, the predations of *Scaphiopus* on alate termites are worthy of note. The most desert-adapted species in North America, *S. couchii*, relies on termites for about 70 percent of its diet (Dimmitt and Ruibal 1980b).

REPTILES

The most numerous, and among the best-adapted, desert vertebrates are the reptiles. In many ways these animals are preadapted for life in arid environments. Their dry, scaly skin, devoid of sweat glands, and their uricotelic method of excretion ensure that water loss through these two main avenues is reduced to a minimum. Three groups of reptiles have solved the problems of living in hot deserts: the tortoises, snakes, and lizards. Desert tortoises are herbivores and are considered later in this chapter.

70   Desert Soil Fauna

Snakes are essentially surface predators, and their depredations on populations of soil-dwelling fauna are as yet incompletely known; in any event the impact of snakes on the functioning of the desert soil system is likely to be minimal. In contrast many desert lizards, which can be broadly classified as insectivores, prey on ants, termites, beetles, spiders, scorpions, and other arthropods. These include members of the genera *Sceloporus, Urosaurus, Callisaurus, Cnemidophorus*, and *Amphibolurus*. More specifically, the horned lizard *Phrynosoma modestum* (Fig. 3.3) is a major predator of honey-pot ants (*Myrmecocystus* spp.) in the Chihuahuan desert. This predatory reptile, like its insect prey, forages in shaded locations during the daytime. However, it is an opportunist, rather than a specialist, as far as its diet is concerned and will switch to other prey items, particularly during the rainy season when the diversity of potential prey species increases (Shaffer and Whitford 1981). Diet switching is common among predatory desert lizards, and on occasion plant material may also be taken. A good example of this is provided by the sand-diving lizard *Aporosaura anchietae* of the Namib, which feeds on kelp flies in coastal areas but has to resort to windblown detritus, mainly grass seed, in inland dunes (Louw 1972). Some desert lizards show definite food preferences, however, and ants and termites figure prominently in their diets. The horned lizard *Phrynosoma cornutum*, for example, feeds al-

Fig. 3.3. The horned "toad" *Phrynosoma* well camouflaged on a Chihuahuan desert soil.

most exclusively on harvester ants of the genus *Pogonomyrmex* (Whitford and Bryant 1979). A predilection for ants is also shown by the lacertid *Eremias* spp. of southwest Africa and the agamid *Moloch horridus* of Australia. Members of the genus *Eremias* also specialize in feeding on termites, as do species of *Ichnotropis, Palmatogecko, Cnemidophorus,* and, in the Kalahari desert, skinks belonging to the genus *Typhlosaurus,* which are markedly fossorial.

A few desert lizards are primarily herbivorous: members of the Old World agamid genus *Uromastix* and the New World iguanid *Dipsosaurus* (Fig. 3.4) and *Sauromalus* are examples. However, as the detailed review provided by Reichman, Prakash, and Roig (1979) indicates, these desert vertebrates are, in the main, secondary consumers, although the introduction of herbivory at this point provides a convenient link between what has gone before and what is to follow.

Desert tortoises (turtles) can be regarded as more committed herbivores than the lizards we have just been considering, despite published statements (Nichols 1953) that captive specimens will feed on snails. The North American *Gopherus agassizi* feeds on succulents and grasses; it is an open-area grazer that can exploit sites where food is relatively abundant. As a consequence it does not store food material in the form of body fat to any great extent

Fig. 3.4. The desert lizard *Dipsosaurus* near Tucson, Arizona.

72 Desert Soil Fauna

(only about 3 percent of the body weight is fat) (Rose 1980). In the same southwestern deserts of the United States *Terrapene* also feeds in open areas and produces only a shallow burrow system, while *Kinosternon sonoriense* is deep burrowing. This latter species is omnivorous and, unlike *Gopherus agassizi*, stores food material in the form of fat to the extent that this storage product can account for up to a third of the body weight. Clearly these desert testudines have evolved at least two, possibly more, approaches to the problems of survival in hot deserts.

RODENTS

These small mammals are widespread in deserts, and examples of them in North America include the kangaroo rats (*Dipodomys* spp.), the pack rat (*Neotoma* spp.), and the pocket mice (*Perognathus* spp.), among others (Figs. 3.5 and 3.6). According to Eisenberg (1975), the rodent fauna of the deserts of North America ranks alongside that of Asia in being the most adapted and in having the longest evolutionary history of any desert rodent fauna in the world. The rodents of the African deserts have evolved to some extent in parallel with those of Asia from which they may be derived. Evidence for this is provided by the distributions of such

Fig. 3.5. Entrances to a *Dipodomys* burrow in the Sonoran desert, southern Arizona.

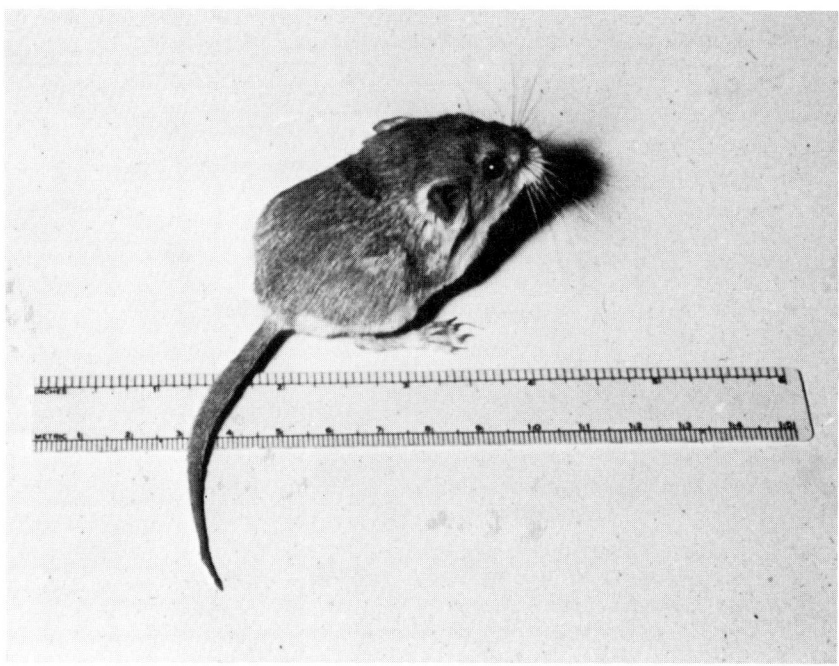

Fig. 3.6. *Microdipodops pallidus*, Esmeralda Co., Nevada. (Photo courtesy of R. E. MacMillen.)

genera as *Acomys, Meriones, Gerbillus,* and *Jaculus,* which extend across both these regions. The Australian desert rodents, exemplified by the murid *Notomys* and *Pseudomys,* show convergences toward other major continental faunas, but their diversity is considerably lower. These convergences illustrate the principle of ecological vicariance, with different genera occupying the same niches in different geographical regions (Table 3.1). As can be appreciated readily from the schema presented in Table 3.1, there is a marked degree of convergent evolution shown by the rodents of North America, Africa, and Asia. The much lower correspondence shown by the Australian desert rodents may be a reflection of the more isolated nature of this fauna or, alternatively, its more recent exposure to the arid conditions for which it is ill-adapted. The latter possibility is supported by the fact that although the arid region of Australia is three times the size of that in North America, it supports only 73 species of mammal, compared with 109 in North America (Morton 1979). Furthermore, the most successful desert rodent faunas, in terms of species richness, are those in which the

TABLE 3.1.
Ecological Vicariants among Genera of Desert Rodents

North America	Africa	Asia	Australia
Citellus (O)	*Paraxerus* (O)	*Citellus* (O) *Meriones* (H, G)	
Peromyscus (O) *Onychomys* (C)	*Acomys* (C)	*Acomys* (C)	
Neotoma (H) *Perognathus* (G)	*Meriones* (H, G) *Gerbillus* (G)	*Gerbillus* (G)	
Dipodomys (G)	*Jaculus* (G)	*Jaculus* (G) *Paradipus* (G) *Dipus* (G) *Scirtopoda* (G)	*Notomys* (G)
Microdipodops (G)		*Salpingotus* (G)	

O = Omnivore
C = Carnivore (including insectivore)
H = Herbivore
G = Granivore
Source: Compiled mainly from data supplied by Eisenberg 1975.

proportion of granivorous species is high. The North American deserts, for example, contain three times as many seed-eating rodents as do the Australian deserts (Morton 1979). But why should this be so? The deserts of both of these geographically distinct regions are of similar ages; so it is not a simple time effect. The answer to this question is in all probability in two related parts.

First, the Australian deserts support three times as many seed-eating ant species (and six times as many granivorous species of birds) as do the North American deserts (Brown, Reichman, and Davidson 1979). This observation prompts the further question: Why are ants (and birds—they are not members of the soil fauna, but their granivorous habit brings them into competition with soil-based animals) the key industry animals, and not the rodents, in Australian deserts? The answer to this question is partly an historical one. The principal granivorous rodents of North America belong to the Heteromyidae, which appeared as early as the Oligocene and could have evolved adaptations to desert conditions as these developed. This early adaptation to arid conditions would allow these rodents to radiate into granivorous niches and would partly exclude ants and birds from this resource (Morton 1979). By contrast rodents did not appear in the Australian deserts until the

Pliocene, by which time these deserts were well established and granivorous niches were already occupied by seed-eating ants and birds. However, this may not be the whole story since, in North American deserts at least, competition between granivorous rodents and ants may be minimized by differences in foraging behavior. Rodents can dig for buried seeds whereas ants cannot. This does not seem to be a mechanism for resource allocation that applies to the Australian situation, since the Muridae of these deserts are not specialized to feed on buried seeds.

Second, Newsome and Corbett (1975) have advanced the suggestion that the low diversity of granivorous rodents in Australian deserts is a reflection of the aridity and climatic variability of these environments. In support of this view they cite the irruptive behavior of Australian desert rodent populations, which is indicative of "r" strategists living in unpredictable environments (see Chapter 10) and which is largely absent from the North American scene. However, aridity and climatic variability in Australian deserts have not resulted in a concomitant reduction in species diversity of herbivores and insectivores, as illustrated in Fig. 3.7; so these factors alone do not provide a convincing explanation for the low diversity of granivorous rodents. On the other hand, an unpredictable environment would give granivorous ants and birds a selective advantage over rodents by virtue of the ability to withstand extended periods of drought (in the case of ants) or the mobility to search out localized food sources (in the case of birds).

In summary the depauperate nature of the Australian desert rodent fauna, in particular its granivorous component, can be attributed to the following factors.

1. It is a relatively recent fauna that appeared in ecosystems where granivorous ants and birds were already well established.

2. It was not versatile enough to exploit subsurface food resources and thereby minimize competition with granivorous ants and birds.

3. It is less able to cope, than granivorous ants and birds, with environments characterized by extreme aridity and the unpredictability of rainfall events.

To this point nothing has been said about the rodent faunas of minor deserts, such as the Monte region of the Argentine, which has a physiognomy reminiscent of the Sonoran and Chihuahuan deserts. Here in South America there are no rodents that can be considered as truly desert adapted (Mares 1975), and this may be attributed to the isolation and recent origin of the Monte region.

76 Desert Soil Fauna

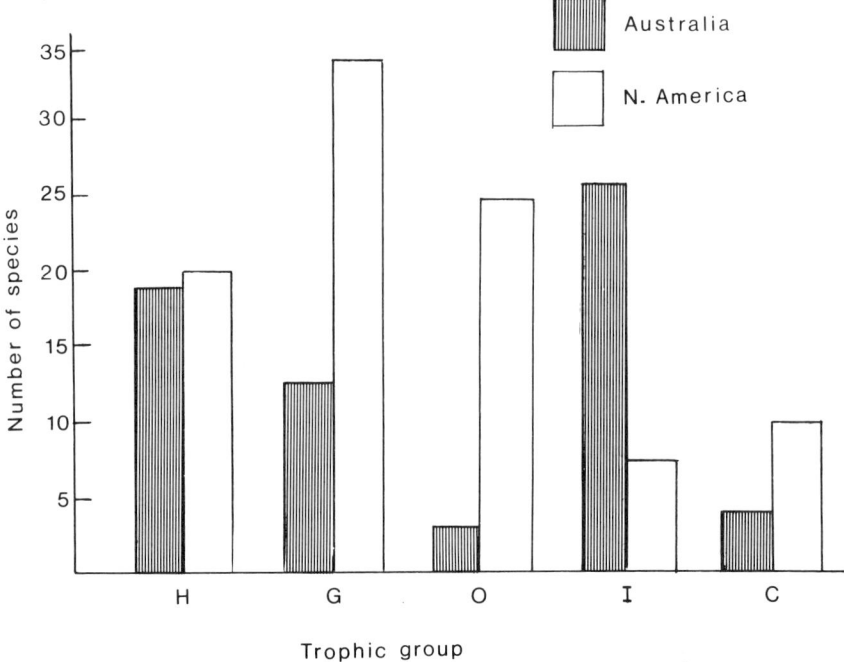

Fig. 3.7. A comparison of the trophic structure of Australian and North American desert mammal faunas: **H** = Herbivore; **G** = Granivore; **O** = Omnivore; **I** = Insectivore; **C** = Carnivore. (After Morton 1979.)

It will already have become apparent, from a perusal of Fig. 3.7, that desert mammals can be classified into a number of different feeding guilds. It will also be evident that a distinction has been made here between carnivory and insectivory, although this distinction will be of doubtful validity to the purist. In practical terms this terminology separates those rodents that take vertebrate prey (carnivores) from those that feed on arthropods (insectivores). There are in fact very few rodents that prey on other vertebrates, in an obligate sense, and this feeding guild, if it exists, can be dismissed summarily from the subsequent discussion. In dealing with the remainder an immediate distinction must be made between the rodents that nest in the soil and forage on its surface and are properly considered in this chapter from those that are completely subterranean and are treated in the next.

Insectivory is uncommon among desert rodents, at least as an obligate feeding habit. Two species of the genus *Onychomys* (*O. leucogaster* and *O. torridus*) are insectivorous in North American deserts, but here the insectivorous niche appears to be occupied, partly at least, by omnivorous rodents (Morton 1979), such as species of *Reithrodontomys* and *Peromyscus*. These species, like *Onychomys* spp., feed mainly on insects when they are available but switch to a diet of seeds during the colder months (Brown, Reichman, and Davidson 1979); this diet switching is reminiscent of certain desert lizards (see earlier). Generalized feeding is virtually absent in desert rodent faunas, except for the murid *Rattus* in Australian deserts (Morton 1979). On the other hand, insectivorous mammals are well represented in Australian desert faunas (see Fig. 3.7), but these are marsupials not rodents, as we have already seen.

Relatively few groups of desert rodents practice herbivory; the North American pack rats (*Neotoma* spp.), which feed on the green leaves of mesquite, sage, and the tissues of *Opuntia*, are examples (Schmidt-Nielsen 1964; Morton 1979; Reichman, Prakash, and Roig 1979). These rodents are able to move through the spiny covering of cacti without injury and have physiological mechanisms for dealing with the high calcium oxalate level in their diet (see Chapter 6). Again, in Old World deserts gerbils of the genus *Meriones* are facultative herbivores, feeding on the vegetative parts of plants during the summer season (Naumov 1975). According to Morton (1979), two genera of Australian desert rodents, *Leporillus* and *Zyzomys*, can be classified as herbivores, but he gives no details of their diet. In Australian deserts, mammalian herbivores are mainly represented by marsupials, notably the kangaroo, *Megaleia rufa*, and *Macropus robustus*, which are not soil-based.

The most specialized desert rodents are the granivores, represented in North America by the genera *Dipodomys*, *Perognathus*, and *Microdipodops*, among others; in Africa and Asia by *Jaculus*, *Meriones*, *Gerbillus*, *Dipus*, and *Paradipus*; while in the Australian deserts the granivorous niche is exploited by the murid *Notomys* and *Pseudomys*. In contrast to the harvester ants that, as we saw in the previous chapter, also occupy an important place in the granivorous niche in hot deserts, these rodents are mainly solitary, multiload foragers, harvesting many seeds in their cheek pouches in a single foray (Fig. 3.8). Foraging areas are usually localized around a permanent burrow system, and these rodents forage both for surface and buried seeds (Brown, Reichman, and Davidson 1979). These are generalizations and inevitably there are exceptions. Members of the Old World genus *Gerbillus* tend to be colonial rather than solitary. Another Old World genus, *Jaculus*, does not restrict foraging activities to the immediate proximity of the burrow (Happold 1975). Two other generalizations that require qualification are that granivorous rodents are all nocturnal foragers and

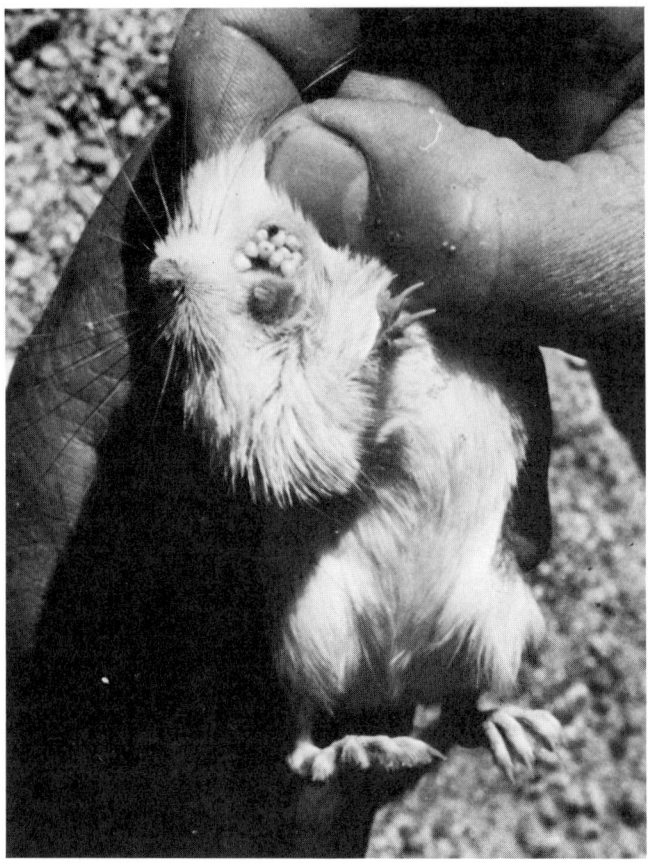

Fig. 3.8. The kangaroo rat *Dipodomys merriami* showing seed-filled cheek pouch, Mohave Co., Arizona. (Photo courtesy of R. E. MacMillen.)

that these animals are compulsive hoarders of food, either in their nests or in shallow caches outside the burrow system (Brown, Reichman, and Davidson 1979). Nocturnalism is the rule among granivorous desert rodents, but there are exceptions; ground squirrels of the genus *Citellus* (= *Spermophilus*) are diurnal in North America, while in the Old World members of the genera *Meriones* and *Jaculus* are semidiurnal or crepuscular. Generally speaking, granivorous rodents in deserts store their food in larders, and this is a device to overcome periods of food stress, which are likely to occur in an unpredictable environment. *Jaculus* spp., however, do not cache food to any marked extent but store food in the form of body fat. Again, *Meriones* spp. hoard seeds erratically. The feeding strategies of all these granivorous rodents are immensely varied, and we will return to this topic in Chapter 8.

Desert rodents, particularly granivores, have a life expectancy of only a few years (Brown, Reichman, and Davidson 1979). Breeding takes place throughout the year in the Australian *Notomys* and *Pseudomys* (MacFarlane 1975) and in the North American *Onychomys leucogaster*, *Peromyscus maniculatus*, and *Reithrodontomys megalotis* (Smith and Jorgensen 1975). Year-round breeding may also occur in the kangaroo rats *Dipodomys merriami*, *D. ordii*, and *D. deserti*, with peaks in the springtime. Other members of this genus have more restricted breeding periods—*D. spectabilis* from January to August, *D. microps* from late January to May. Species of the genus *Neotoma* also show differences in their patterns of breeding behavior, although the discussion of these by Smith and Jorgensen (1975) is, to say the least, confused. These authors have assembled a considerable body of data on the reproductive biology of North American desert rodents, which suggest that few generalizations can be made. Not only are breeding seasons variable in length and timing, but also there are wide variations from species to species in litter sizes and the length of time to puberty. These variations may be more a reflection of the response of a population to local climatic conditions and food availability than to any intrinsic properties of the species. Moreover, these findings must be interpreted with some caution since they incorporate observations on laboratory animals. However, there are some general points that do emerge and that do suggest that reproductive strategies are designed to cope with arid conditions. First, although breeding activity may occur over extended periods, it often peaks at times of the year when there is an active growth of vegetation. This illustrates the "pulse" phenomenon characteristic of hot deserts (see Chapter 5). Second, rodents inhabiting the more arid regions of deserts tend to produce fewer litters per year than rodents living on the margins of deserts. The majority of true desert species give birth to 1 or 2 litters each year, whereas species such as *Reithrodontomys megalotis* and *Baiomys taylori*, which are marginal desert dwellers, may produce as many as 7 and 10 litters per year, respectively. This distinction extends to litter size, with some important exceptions (see below). Litter sizes tend to be lower in those species adapted to truly desertic conditions, compared with their relatives in marginal habitats. For example, *Peromyscus maniculatus* is widespread throughout the United States and, in semidesert sites in Texas and Nevada, can produce up to nine young per litter, with averages of four to five. In contrast the related *P. eremicus*, which is restricted almost entirely to desert habitats, averages three young per litter, with a maximum of five.

The suggestion that emerges from this analysis is that many desert rodents respond to the vagaries of their environment by reducing their reproductive output. The recruitment of relatively few individuals into the population, by reducing litter size and number, has several obvious advantages. First, it may maximize available food resources. Second, it may reduce the amount of milk that needs to be produced by a lactating female and thereby conserve precious water. As Schmidt-Nielsen (1975) points out, this is a topic that deserves further attention in relation to water budgets of desert rodents. Again, reduced litter size will facilitate more effective parental care against predation. Predation is an ever-present threat to rodent populations—rattlesnakes and hawks by day, owls and foxes by night—and at low population densities predation may deplete rodent numbers to a point where they would find it difficult to recover. Rodents have developed subtle mechanisms to counteract this threat, however. For example, the bannertail kangaroo rat, *Dipodomys spectabilis*, does not undertake moonlight excursions on the soil surface during the winter months but may do so in the spring and autumn. In the winter when food supplies are at a minimum, the rewards from foraging are outweighed by the risks of predation, whereas the reverse may be true in the spring and autumn (Lockard and Owings 1974; Schroder 1979). Furthermore, *Dipodomys* and *Microdipodops* species that are truly desertic have enhanced auditory sensitivity, compared with mesic species (Webster and Webster 1980), and this has been shown to be of adaptive value in predator avoidance (see Chapter 9). This is a slight diversion, but a pertinent one. Now back to the main discussion.

Heteromyids, of which members of the genus *Dipodomys* are examples, have the lowest reproductive output of all the desert rodents, but they are followed closely by the sciurids in this regard. Table 3.2 shows that sciurids produce one, less commonly two, litter(s) per year, and this compensates for their relatively large litter sizes. It must be pointed out that although the estimates presented in Table 3.2 are based on data for nondesert sciurids, the available information indicates that they apply equally well to such desert dwellers as *Citellus nelsoni* and *Spermophilus* spp. (Hawbecker 1975; Smith and Jorgensen 1975). In contrast to Hawbecker's (1958) findings, the data given in Table 3.2 suggest high rates of survival in sciurids, and French, Stoddart, and Bobek (1975) attribute this to seasonal dormancy, which occurs in many sciurids, including the desert-dwelling *Spermophilus mexicanus*, *S.*

TABLE 3.2.
Reproductive and Demographic Data for Rodents, with Particular Emphasis on Desert-Dwelling Groups

	Number of Litters (N)	Litter Size (L)	Output (L × N)	Survival (months)	Density ha^{-1} (seasonal/ interseasonal)
Murid	3.30	6.14	20.3	1.82	117.7–15.7
Cricetid	3.06	4.04	12.4	3.65	15.4–10.6
Sciurid	1.44	5.53	8.0	12.53	6.7–11.8
Heteromyid	2.29	3.05	7.0	9.28	8.6–8.5

Source: Compiled from data provided by French, Stoddart, and Bobek 1975.

tereticaudus, and *S. townsendi* (Smith and Jorgensen 1975). Low reproductive output and high rates of survival interact to produce low but stable population densities, and these are characteristic of both the sciurids and the heteromyids.

Cricetids, such as members of the genera *Gerbillus*, *Meriones*, *Onychomys*, *Peromyscus*, *Reithrodontomys*, and *Tatera*, have higher reproductive rates on average than sciurids and heteromyids (Table 3.2), and this may be a device to counteract the modest survival rate of these rodents. However, like the sciurids and heteromyids that are also slow breeders, cricetids occur in stable, low to moderate densities. By contrast the murids have a high reproductive rate and low probabilities of survival, which combine to produce marked seasonal fluctuations in population densities (Table 3.2). As desert dwellers, murids only come into prominence in Australia, and here *Mus, Rattus, Notomys*, and *Pseudomys* species do not go into dormancy but breed for most of the year. These volatile populations produce explosions in numbers from time to time (Newsome and Corbett 1975), as might be expected from their high reproductive and turnover rates (French, Stoddart, and Bobek 1975).

Although much of the desert soil fauna is invertebrate in character, the rodents have been given a special place in this book for a number of reasons. First because the soil plays an important part in their survival. The rodents we have been considering in this chapter are as much a part of the soil fauna as the surface-foraging millipedes, centipedes, scorpions, termites, beetles, and ants. Second, because many of them occupy the important granivorous niche in hot deserts. Third, because their feeding activities impinge on those of other members of the soil fauna, such as the granivor-

82 Desert Soil Fauna

ous ants (Brown and Davidson 1977; Reichman 1979), and there is need for a much greater exchange of information between desert mammalogists and soil zoologists. Finally, because the strategies for survival employed by desert rodents have been explored in much greater detail than have those of desert soil invertebrates, and much of what relates to the former may be relevant to the latter. This topic is taken up in greater detail in Chapters 9 and 10. Figure 3.9 will suffice for the moment to summarize the three different types of survival strategy exhibited by the rodents we have been considering.

LAGOMORPHS

Jackrabbits (*Lepus* spp.) and cottontails (*Sylvilagus auduboni*) in North American deserts have perhaps less of a claim to be

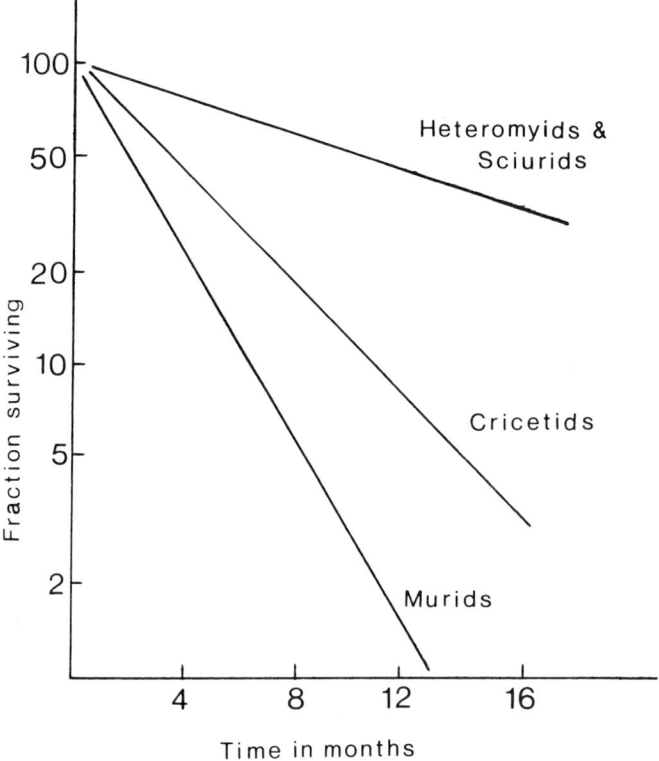

Fig. 3.9. Survival curves for three groups of desert rodents. (From French, Stoddart, and Bobek 1975.)

considered as soil animals than any of the other groups treated in this book. Their activities are very much confined to above-ground situations where they browse on shrubs and herbs (Chew and Chew 1970; Reichman, Prakash, and Roig 1979). However, much of plant material consumed by these herbivores may eventually find its way, in the form of fecal pellets, onto, or below, the soil surface. Here it may provide a food base for tenebrionid beetles, for example, and thereby form a link between aerial and subterranean parts of the desert ecosystem. Such a link has been forged through the feeding activities of desert termites on cattle dung (see Chapter 2). In view of their often high densities it could be expected that desert-dwelling lagomorphs would make a sizable contribution to this dung pool, but no data are at hand to document this. This is an area of research that could yield fruitful results, particularly in the context of the overall functioning of desert ecosystems—yet another area that needs the combined attention of mammalogist and soil zoologist.

The importance of this cooperation should not be underestimated. Studies on organic decomposition currently being carried out in the deserts of the American Southwest have identified creosote bush litter and, to a lesser extent, the discarded leaves of mesquite as important sources of organic input to soils (Santos, DePree, and Whitford 1978). In this context it is perhaps significant to note that lagomorphs browse on fresh leaf material both of creosote bush and mesquite (Chew and Chew 1970; Shoemaker, Nagy, and Costa 1974; Turkowski and Reynolds 1974) and that they frequently discard partly masticated bolus into sheltered soil depressions. This semidigested material evidently decomposes in the soil at a faster rate than "untreated" material (Whitford 1980: personal communication). Here again is a point of contact between two compartments of the desert ecosystem—an exciting prospect for future research.

4
The Subterranean Fauna

The distinction between the true soil fauna and the surface-active fauna is often an arbitrary one, and it would be unwise perhaps to be too inflexible in identifying these two ecological divisions. As we have seen in the previous chapters, the surface-active fauna of hot deserts is dominated by animal groups of relatively large body size. When attention is focused on the animals living in more permanent, and more intimate, contact with the soil environment, then—apart from the subterranean mammals—the scale of body size diminishes by an order of magnitude or more. Here we are in the realm of the microscopic animals, the microfauna and the mesofauna, as ecologists like to call them. These animals, together with the living and dead plant material that forms an important part of their environment, represent a subsystem of the larger desert ecosystem that encompasses the surface-active organisms, with which we have already dealt, and the aerial, or epigeal, subsystem, which lies outside the scope of this book.

Like its surface-active counterpart the subterranean community has a more or less well-defined feeding, or trophic, structure. The energy base is provided by aerial parts of desert plants: leaves, shoots, and woody stems that die back or that are discarded in some other way (for example, as rejecta) to become partly or completely buried in the soil. This burial process is facilitated by the transport of dead plant material in surface runoff down arroyos and wadis during the rainy season (see Chapter 5). This buried organic litter is decomposed primarily by bacteria and fungi but (as will be seen later on) the activities of these microorganisms are very much controlled by the desert soil fauna. Among this fauna are groups that can be considered as primary consumers: these are the animals that feed mainly on plant detritus or directly on the microflora. Included here are various nematodes, myriapods, Collembola, Diplura, higher insects, mites, and mammals. Earthworms are generally considered to be absent from arid soils, but recent observations by Perseu Santos and Walter Whitford (1979: personal communication) in North America and Jo Springett (1979: personal communication) in Australia have demonstrated that members of the Enchytraeidae (pot-worms) may colonize buried litter in large numbers in hot deserts.

Subterranean secondary consumers in hot deserts are represented mainly by arachnids (mites, pseudoscorpions, and spiders), together with insectivorous mammals. However, the term *predator* has also been applied to animals that feed on bacteria, such as various nematodes, Collembola, and oribatid mites. To pursue this line of thought further would involve a descent into the realm of semantics, which is not our purpose. Instead, we can proceed to examine the biological characteristics of the various animal groups, starting with the nematodes.

NEMATODES

This group of invertebrates has a worldwide distribution in soils, and the centers of population density are usually concentrated around the rooting systems of plants—the zone known as the rhizosphere. Freckman and Mankau (1977) found this to be true of soil-inhabiting nematodes in Rock Valley, Nevada, a Mojave desert site. Here the dominant vegetation is composed of the shrubs *Larrea tridentata, Ambrosia* (= *Franseria*) *dumosa, Krameria parvifolia,* and *Lycium andersonii.* Soil was sampled for nematodes at depths down to 30 cm from the shrub base and at a distance of one and three shrub radii from this base (Fig. 4.1). Over a 2-year period nearly 3,000 soil samples were analyzed, and the results showed a significant decline in nematode numbers with depth and distance from the shrub base.

Soil nematodes are not very easy to classify taxonomically, but they can be assigned, a little more readily, to various feeding groups. Several classifications based on feeding habit have been proposed from time to time, and these are summarized by Yeates

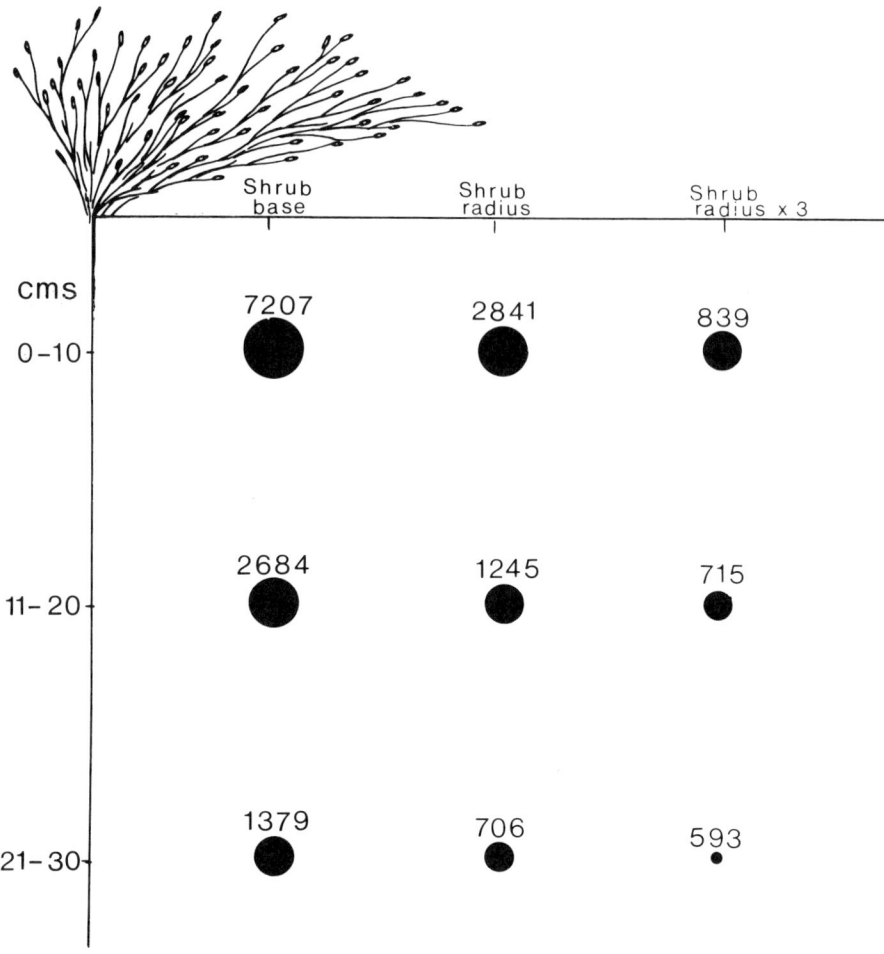

Fig. 4.1. Spatial distribution of soil nematodes in relation to desert shrubs at Rock Valley, Nevada (numbers per 500 cm^3 of soil). (After Freckman and Mankau 1977).

(1971). The virtue of this type of classificatory scheme is that it lays emphasis on ecological function. The trophic structure of the nematode fauna in different sampling sites can be compared and shifts in the relative proportions of the different feeding categories may be indicative of differences in the ecological processes they promote. For example, the predominance of fungivorous nematodes over bacteriophages in a particular site would suggest that decomposition of organic substrates is proceeding along fungal, rather than bacterial, pathways.

Freckman and Mankau (1977) identified four trophic groups of nematodes at Rock Valley, namely: (1) fungal feeders, such as *Aphelenchus avenae, Aphelenchoides* sp., and *Ditylenchus* sp.; (2) microbial feeders, mainly members of the Cephalobidae; (3) omnivore/predators (mostly Dorylaimina); and (4) plant parasites. Microbial feeders occurred in the greatest numbers in virtually all of the sampling stations, followed by omnivore/predators, fungal feeders, and plant parasites. The numbers of microbial feeders and omnivore/predators showed a significant decrease with depth, however, and this may be attributed to the dilution of bacterial populations with increasing distance from the rhizosphere.

The relative proportions of different trophic groups under the various shrub species deserves further scrutiny, for this may provide an insight into the nature of decomposition processes in hot desert soils. Freckman and Mankau (1977) found that plant-parasitic nematodes were more numerous under *Lycium andersonii* and *Larrea tridentata* than under *Ambrosia* and *Krameria*. Omnivore/predators did not show any particular preference for plant species, as might be expected, since their densities are likely to be regulated by the availability of prey and by a variety of organic substrates. The other associations suggest, however, that bacteria may be more important as agents of decomposition than fungi in *Ambrosia* sites, whereas the reverse may well be the case under *Krameria*. Again, the prevalence of plant parasites under *Lycium* and *Larrea* indicates that herbivory, rather than detritivory, is the key industry here and that grazing food chains form an important structural base for this type of community. These are only indications; they require further documentation.

The mean annual density of nematodes at Rock Valley was determined by Diana Freckman (1980: personal communication) to be 1.24×10^6 m^{-2}. This is appreciably higher than the estimate from the vegetated cold desert (Arctic) soil investigated by Chernov, Striganova, and Ananjeva (1977) but appreciably lower than many estimates obtained from cool, moist temperate soils (Table 4.1). Even the figure of 1.24 million for a hot desert soil may give an exaggerated impression of nematode activity here since these animals can enter a quiescent, nonmetabolic state, termed *anhydrobiosis*, during periods of drought (Figs. 4.2 and 4.3). This state of suspended animation, which is described in some detail in Chapter 6, is induced at least in part by exposure to a very slow desiccation process (Crowe and Madin 1975; Evans and Perry 1976). At Rock Valley nematodes of all trophic groups went into an anhydrobiotic state during the period of the summer drought (April to October), although they could be reactivated by sudden rainfall events

TABLE 4.1.
Population Densities of Soil Nematodes in Cool Temperate and Desert System

Habitat	Sample Depth (cm)	Numbers ($\times 10^6$ m^{-2})	Author
Beech floor	25	12.1	Volz 1951
Oak floor	25	29.8	Volz 1951
Hot desert	30	1.24	Diana Freckman 1980: personal communication

Source: Compiled by the author.

during the month of August. This is evidence of opportunism on the part of the soil nematode fauna, and we will refer to it again later in this book.

The anhydrobiotic condition presents a problem to the investigator who wishes to obtain an accurate estimate of the number of active nematodes in a particular site at any given time. The most widely used methods for extracting nematodes from soils involve elutriation techniques (Seinhorst 1962) or wet funnels of the Baermann type (Nielsen 1949). These have the disadvantage, as far as extraction of desert soil nematodes is concerned, of using water as the extractant. Rehydration of the nematodes occurs during the extraction, and any anhydrobiotic individuals that are present in the soil at the time of sampling appear in the extract in an active state. Further, extraction in water encourages eggs present in the soil to hatch, and this inflates the numbers of nematodes recovered. Both of these effects will produce overestimates of the numbers of active nematodes in soils at the time of sampling.

To overcome this problem, a sugar flotation-seiving technique has been used that allows nematodes to be recovered from desert soils in their natural state, anhydrobiotic or active (Freckman, Kaplan, and Van Gundy 1977); the specimens depicted in Figs. 4.2 and 4.3 were obtained by this method. The efficiency of this technique for anhydrobiotic nematodes is on the order of 80 to 95 percent when extracted in 1.25 M or 1.5 M sucrose solutions.

ENCHYTRAEIDS

These minute, whitish worms are generally associated with soils rich in moist, decomposing organic matter (Wallwork 1970) where they may feed selectively on fungi (O'Connor 1967). It is of

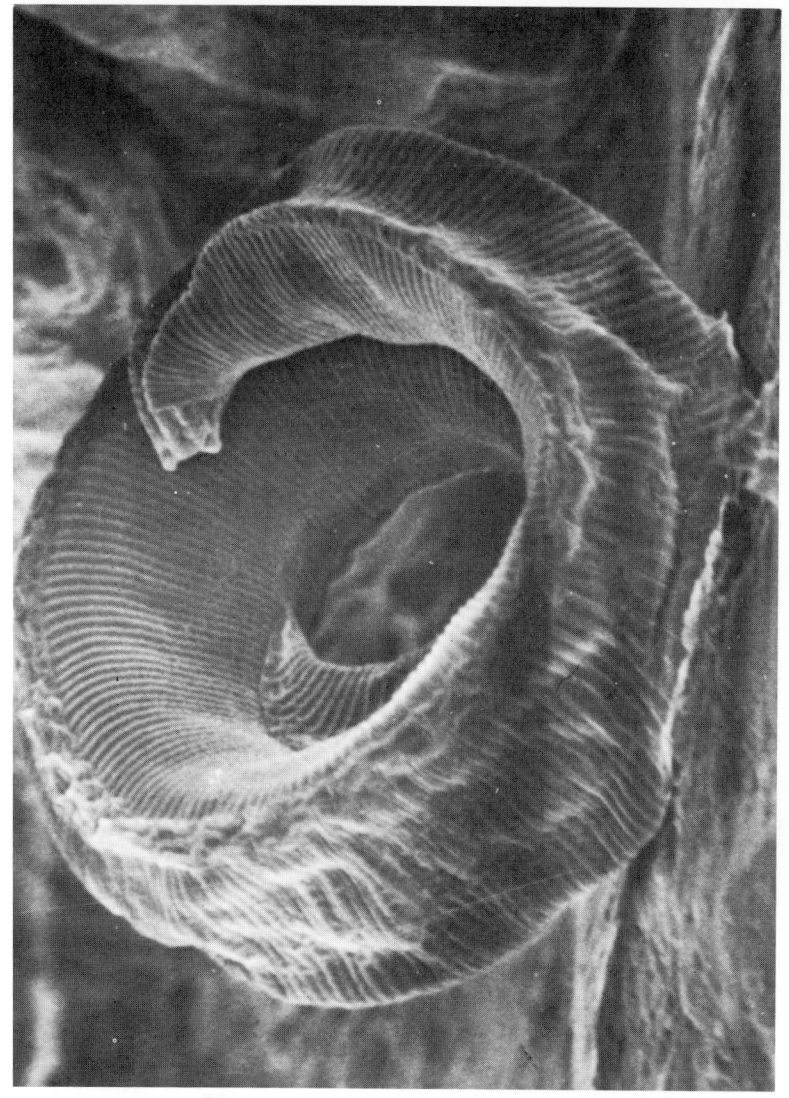

Fig. 4.2. The bacterial-feeding nematode *Acrobeloides* sp. in an anhydrobiotic state x 1288. (Photo courtesy of Diana Freckman.)

90 Desert Soil Fauna

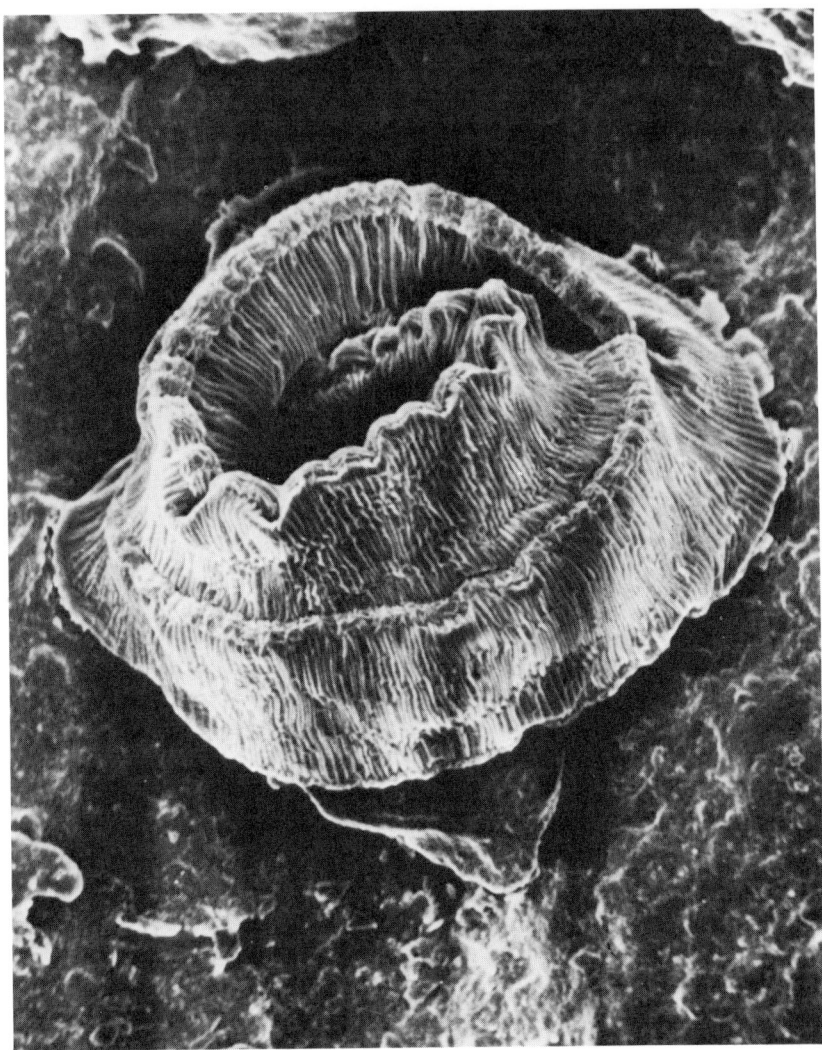

Fig. 4.3. Anhydrobiosis in the fungivorous nematode, *Aphelenchus avenae* x 1104. (Photo courtesy of Diana Freckman.)

some interest to note, therefore, that they have been recovered from arid sites in North America and Australia. Santos and Whitford (1979: personal communication) recovered enchytraeids in large numbers from litter bags (containing mixed desert litter) that had been buried beneath the soil surface for a year. These organic islands also develop into moisture islands when covered with soil and, as such, will attract enchytraeids. On the other hand, these worms must originate from sources outside these artificial environments; Santos and Whitford suggest that they may survive in natural islands of buried litter, although this remains to be verified. Sites that would repay further scrutiny, in this regard, are the organic soils under large shrubs and trees on high bajada sites. Populations of enchytraeids living in these relatively mesic environments could provide a constant supply of colonists for buried litter at lower elevations through the agency of rainwater runoff. The extent to which such runoff moves small soil animals, such as enchytraeids, from high to low elevations in hot deserts is not known; it is worthy of study, and this would not be difficult, providing the investigator does not mind getting wet! Be this as it may, one might expect to find that the enchytraeids that occur in hot deserts will be the most drought resistant of their kind; certainly, this is the case with the Collembola and Acari, as we will see a little later. Enchytraeids present taxonomic problems to the nonspecialist that are, to say the least, difficult, and we know virtually nothing about the identity of the forms that inhabit desert soils. Jo Springett (1979: personal communication) has tentatively identified some enchytraeids, collected from arid soils in Australia, with the genus *Marionina*, and if this is confirmed it would not be surprising. Members of this genus are noted for their ecological plasticity, and the selective pressures of the desert soil environment could be expected to work in their favor.

MYRIAPODS

The most familiar myriapodous arthropods are the centipedes and the millipedes. Many of these occur in hot deserts as surface-active forms, and they have been considered in this context in Chapter 2. Subterranean centipedes belonging to the Geophilomorpha also occur in hot deserts, as do the tufted millipedes of the genus *Polyxenus*. In addition there is another group of myriapods that occurs in hot deserts and is completely subterranean, namely, the Pauropoda.

Geophilomorph centipedes apparently possess a lipid, waterproofing layer in the epicuticle, a feature that would preadapt them for life in an arid environment. However, they have no efficient spiracular closing mechanism (Blower 1955), and they are therefore consigned to a completely subterranean existence. Their occurrence in desert soils is not well documented, and it would appear to be very sporadic. In a survey of the soil fauna in a range of habitats in the Chihuahuan desert of New Mexico (Wallwork, in preparation), geophilomorphs were collected only from organic debris beneath Apache plume (*Fallugia paradoxa*). Here they occurred in small numbers and were mainly concentrated in the litter layer. *Geophilus* sp. has also been collected in low numbers from coastal Mediterranean desert soils in Egypt (Ghabbour, Mikhail, and Rizk 1977). Nevertheless, this group of myriapods deserves further attention, for it may be more widespread in hot desert soils than the above observations indicate, particularly in organic substrates at high elevations. Furthermore, as a predator of such key industry primary consumers as Collembola and juvenile mites, the geophilomorph centipede may be an important regulator of population size in these saprophagous groups. Clearly, this is an area in which further research is needed.

Little can be said at this stage about the occurrence of polyxenid millipedes in hot desert soils. Wallwork (in preparation) recovered a single specimen from juniper litter in a high bajada site in the Chihuahuan desert. This is a rare group of millipedes, and although its occurrence in hot desert soil is interesting, its contribution to the functioning of desert soil ecosystems is undoubtedly minimal.

Very little is known also about the biology of pauropods. They do not appear to have the same kind of ubiquity as the enchytraeids, collembolans, and mites. Published records (Starling 1944; Lawrence 1953; Cloudsley-Thompson 1958) indicate that they favor the moist, organic conditions provided by decomposing forest litter where food material in the form of fungi may be abundant. However, large numbers of pauropods have been collected from organic soils under *Prosopis* and *Fallugia* growing on a high bajada site in the Chihuahuan desert (Wallwork, in preparation). These collections, which were taken immediately after heavy summer rains, contained larval as well as adult stages, indicating well-established breeding populations in these sites (see also Chapter 5). Working in this same area, Yossi Steinberger and Walter Whitford (1980: personal communication) have recorded pauropods from a variety of habitats. As far as is known. there are no other published records of the occurrence of pauropods in desert soils. They may represent, like the enchytraeids, a mesic element in the desert soil fauna, and a study of their ecology, particularly in relation to survival strategies and their impact on decomposition processes, promises to yield interesting results.

COLLEMBOLA

Springtails occur in soils throughout the world, from the cold polar deserts to the equatorial forests; so their presence in hot deserts comes as no surprise. In a survey of Collembola in arid soils of the Sudan and southern Australia, P. Greenslade (in press) identified a total of 19 genera belonging to 6 families. *Xenylla, Brachystomella, Pseudachorutes, Folsomides, Proisotoma, Cryptopygus,* and *Sphaeridia* were common to both of these geographically distinct areas, although they were represented by different species in each. Again, this is to be expected since several of these genera have a wide geographical distribution. *Xenylla* and *Folsomides* occur in the Mojave desert (Wallwork 1972a), while *Cryptopygus* is widely distributed in the Maritime and sub-Antarctic zones of the south polar region (Tilbrook 1967).

Many Collembola are detritivores, ingesting particles of decaying plant material; others may feed preferentially on fungi and bacteria (Christiansen 1964). Still others are known to be herbivores, and these include species of *Onychiurus*, which attack the living roots of plants, and members of the symphypleone Sminthuridae, which can feed in the aerial parts of the vegetation, on leaves, pollen, and algae. As will become evident in Chapter 6, sminthurids are adapted to arid conditions by virtue of their possession of a waterproofed cuticle, and they occur quite commonly in soils of hot deserts, particularly those of an organic nature. What they feed on here is a matter for conjecture, but their proven ability to feed on living plant material may be an advantage in environments where alternative food sources (such as dead plant material) may be dry and unpalatable for most of the year.

The habit of feeding on bacteria adopted by various collembolan species deserves further attention since bacteria are evidently important promoters of organic decomposition in hot deserts (see below). Members of the genera *Folsomides* and *Neanura* (Fig. 4.4), which occur in North American deserts, may fall into this category. Such bacteriophages may regulate, indirectly through their feeding activities, decomposition rates; this hypothesis remains to be tested.

It is well established (see Wallwork 1970) that the Collembola, as a group, comprises a number of life forms, each of which can be identified with a specific habitat, as it is expressed in the vertical dimension. Species that are habitually associated with aboveground (epigeal) habitats are usually highly mobile, with well-developed springing organs, strongly pigmented, and possess well-

94 Desert Soil Fauna

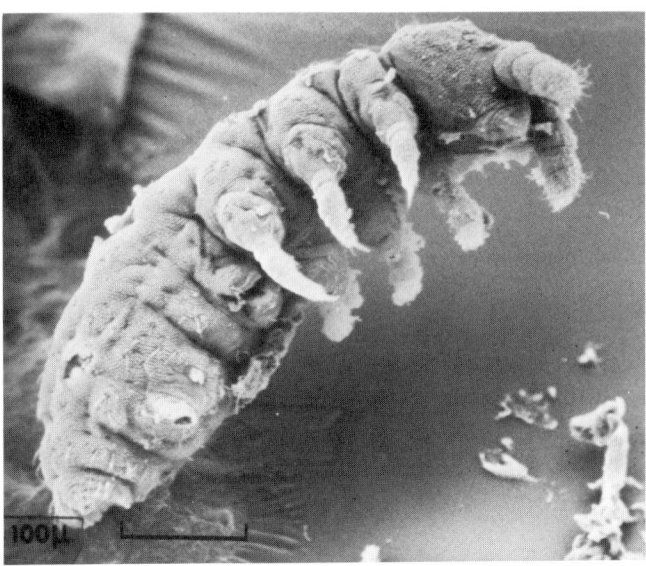

Fig. 4.4. A neanurid collembolan from the Chihuahuan desert of New Mexico. Bar scale = 100 μ. (Photo by Chris Walker.)

developed eyes and antennae and a waterproofed cuticle. All of these characteristics are present in the Sminthuridae, and, as we have just noted, this is a group that has established itself in hot deserts and evidently is able to exploit epigeal habitats here. P. Greenslade (in press), for example, has reported collecting *Deuterosminthurus* sp. and *Corynephoria* sp. in sweeps from vegetation in the Sudan and south Australian deserts, respectively, during the daytime when air temperatures exceeded 30°C. In both of these genera the tracheal system is well developed, and the cuticle bears tubercles that may aid in the retention of a lipid layer and promote a "plastron" effect (Noble-Nesbitt 1963). Not all sminthurids are epigeal in habit, however. Some of the smaller-sized species occur in litter and, indeed, in the underlying mineral soil, and they have been recovered from these microsites under *Prosopis* and *Fallugia* in the Chihuahuan desert (Wallwork, in preparation).

The Collembola that live in the surface litter are usually referred to as hemiedaphic. These are species in which the eyes and cuticular pigmentation are usually well developed, but the antennae and springing organs show some variation in their development. Many members of the family Entomobryidae can be assigned to this ecological grouping, such as *Lepidocyrtus*, *Seira*, and *Acanthocyrtus*, which have been recorded from desert soils (P.

Greenslade, in press). This family also includes the genus *Pseudosinella* whose members are not adapted to dry environments but, nevertheless, have been recorded from Australian deserts (P. Greenslade, in press). The species in question has reduced ocelli and a lack of pigmentation, which are features of a deep-soil dweller. Two other families that are mainly hemiedaphic and that occur in desert soils are the Hypogastruridae and Isotomidae, the former represented by *Xenylla* and the latter by *Folsomides, Proisotoma*, and *Cryptopygus* (see above). However, *Proisotoma brisbanensis* occurs rather deep in the soil and can hardly be considered as hemiedaphic. The same is true of species belonging to the hypogastrurid genus *Acherontiella* in south Australia (P. Greenslade, in press). These inhabitants of the deeper soil layers are considered to be euedaphic and are often characterized by a lack of pigmentation, short antennae, and springing organs and ocelli reduced or absent. This euedaphic grouping embraces, in desert soils as elsewhere, members of the family Onychiuridae.

Life in the deeper horizons of the soil profile is restricted to the pore spaces between soil particles. Here light is not an environmental factor of any importance, so that body pigmentation and a well-developed sense of sight are of little consequence. Movement in the restricted spaces would be impeded by the possession of long antennae; by a similar token, well-developed springing organs would be of little value. The euedaphic life form, like its hemiedaphic and epigeal counterparts, is well suited to its environment, but unlike the surface dwellers its members employ an avoidance, rather than a tolerance, survival strategy. The epigeal Sminthuridae have tolerance mechanisms that are designed to curtail the loss of water from the body in dry environments, as we have already seen. The hemiedaphic isotomid genus *Folsomides* has the ability to assume a wilted state during drought conditions (see Chapter 6), akin to anhydrobiosis in nematodes (see earlier). Drought resistance is also shown by members of the genera *Brachystomella* and *Setanodosa*, which are desertic representatives of the family Neanuridae. These neanurids are hemiedaphic forms, but although they are pigmented as befits their life-style, their mobility is reduced by the absence of a springing organ. This sedentary mode of life increases their vulnerability to predators, but this is offset by their possession of cuticular spines and possibly their ability to produce repugnatorial secretions.

P. Greenslade (in press) has been at pains to point out that some collembolan populations respond immediately to rainfall events. She cites examples from the south Australian desert, and that of the Sudan, of species that show increased activity after rainfall events. Anhydrobiotic species, such as members of the genera *Folsomides* and *Brachystomella*, respond to the presence of environmental moisture very rapidly by resuming their normal appearance, while newly hatched immatures and adults of *Setanodosa* sp. were active immediately after April rains. Members of all three of these genera have, in addition to an inactive adult stage, drought-resistant eggs that can carry the species through from the end of one rainy season to the beginning of the next. On the other hand, P. Greenslade (in press) notes that one group of *Folsomides* species avoids drought conditions at the surface by living in the deeper soil layers, and this avoidance allows them to be active continually throughout the year. The same is true of other euedaphic groups, such as *Onychiurus* and *Acherontiella* species. Year-round activity is not confined to these deep dwellers, however; it occurs in the hemiedaphic entomobryids *Lepidocyrtus*, *Seira*, and *Acanthocyrtus* and in the epigeal sminthurids *Corynephoria* and *Prorastriopes*. In many of these cases continuous activity through the seasons is made possible by the possession of physiological tolerance mechanisms already mentioned.

DIPLURA

These little-known arthropods had not previously been recorded from hot desert soils, as far as is known, until they were found in samples of litter and mineral soil taken from beneath *Fallugia*, *Prosopis*, and *Quercus* in a bajada site in New Mexico (Wallwork, in preparation). All of the specimens collected were members of the Japygidae and, as far as could be ascertained, belonged to a single genus. They showed no particular preference for litter or mineral layers, being equally common in both, and they were also recovered from leaf litter under juniper, creosote bush, and black grama grass on the high bajada. These are predaceous arthropods, and, in the black grama site at least, their numbers would suggest that they are an important group of carnivores. They merit further attention, particularly in the context of predator/prey systems in hot deserts.

INSECTS

Already in Chapter 2 we have considered many of the desert insects that are associated with the soil but that forage on its surface from time to time. These are conventionally considered to be periodic members of the soil fauna (Wallwork 1970). We can now turn to those groups of insects that belong to the permanent soil fauna—those that are more intimately associated with the subterranean habitat during their entire life-span. Relatively few groups of insects fall into this category in hot deserts; the primitively wingless bristletails (Thysanura), thrips (Thysanoptera), plant lice (Psocoptera), and some of the termites (Isoptera) are examples. Some, at least, of these may appear on the soil surface intermitently, but they differ from the members of the periodic fauna in that they feed for the most part on organic debris in the litter horizon rather than on the aerial parts of the vegetation. In some instances this distinction may be rather arbitrary because of diet switching, but there is no ambivalence between these two categories, on the one hand, and temporary soil fauna, on the other. The latter comprise those insects that are subterranean for only a part of their life cycle: usually the vulnerable egg and juvenile stages. Once development has been completed, the winged adults leave the soil and live in the aerial environment. Diptera larvae belong to this ecological grouping, and, to judge from their occurrence in leaf litter under *Atriplex, Quercus, Juniperus, Prosopis*, and *Fallugia* in southern New Mexico (Wallwork, in preparation), they are widespread in deserts. The voracious feeding habits of these insects would suggest that they may be important agents in the fragmentation of organic litter, but little attention has been paid to this so far. The same is true of the beetle larvae that are common in desert litter; many of these are undoubtedly the immatures of periodic species, but others are temporary members of the soil fauna. Examples of the latter are the case-building larvae of chrysomelid beetles belonging to the genus *Cryptocephalus*, which occur in *Atriplex* litter at White Sands, New Mexico. These periodic, temporary, and permanent categories of desert soil fauna are compared schematically in Fig. 4.5.

Thysanurans are generally noted for their ability to colonize dry environments. In cool temperate regions members of the genus *Machilis* are abundant in rock crevices of the marine littoral zone, whereas the silverfish *Lepisma*, and the firebrat *Thermobia* are commonly associated with buildings and can be encountered in bathrooms, cellars, and bakehouses. Evidently, the group as a whole has good powers of survival in atmospheres below saturation level and in this sense may be considered to be preadapted for life in deserts. As Crawford (1979b) points out, lepismatid thysanurans are widely distributed in arid regions. They may be restricted during the day to cryptic habitats under rocks, dead vegetation, or dung; some species, on the other hand, seek the shelter and protec-

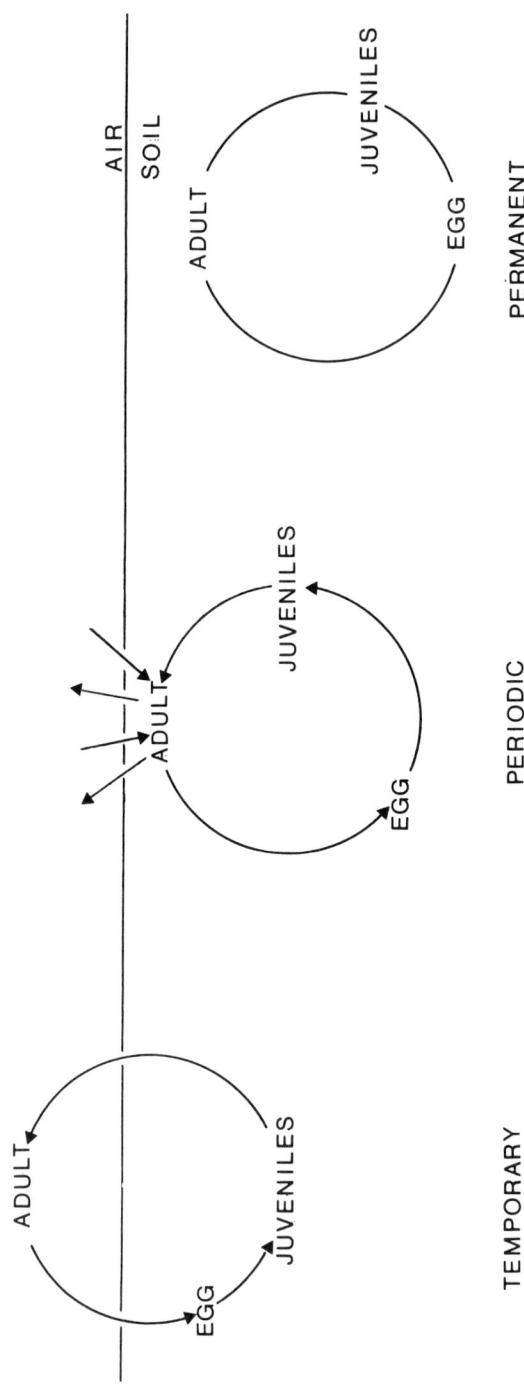

Fig. 4.5. Ecological types of desert soil insects defined according to their degree of presence in the soil. (Adapted from Wallwork 1970.)

tion of ant and termite nests. However, thysanurans may be active on the soil surface at night, even in sand dune areas, and this activity is made possible, at least in the case of *Ctenolepisma terebrans* of the Namib, by an unusual physiological mechanism for gaining moisture from the desert air (see Chapter 6).

Thysanopterans (thrips) are often associated with the aerial parts of the vegetation in cool, moist temperate regions, although they may also be recovered from leaf litter, usually as nymphs. This suggests that these insects are temporary members of the soil fauna in these regions. In the Chihuahuan desert, however, adult thrips have been extracted from litter under *Fallugia*, *Quercus*, *Larrea*, and *Bouteloua*, which indicates that here they may constitute part of the permanent soil fauna. This is possibly an expression of a phenomenon, noted by Wallwork (1976), of a vertical compression of the faunal profile downward toward the soil in extreme environments.

Thysanurans and thysanopterans are probably detritivores in hot desert soils, although the low densities in which they occur would hardly seem to qualify them as key industry animals in the events associated with the decomposition of organic material in such soils. Psocopterans cannot be dismissed so lightly, however. Like the bristletails and thrips, these insects are globally distributed in dry habitats but, unlike these other two groups, psocopterans can be locally abundant in desert soils. The data provided in Table 4.2 indicate that psocopterans are more abundant in the litter layer, as opposed to the mineral, in a range of desert organic soils. These data are based on a limited number of samples, but they substantiate the observations of Perseu Santos and Walter Whitford (1979: personal communication) that Psocoptera consistently feature in faunal extracts from desert litter. These workers noted that liposcelids appeared during the late stage of litter decomposition. Little is known of the feeding habits of desert psocopterans; liposcelids may be fungivorous, but this is an area that deserves further study, particularly in relation to decomposition processes.

In Chapter 2 a distinction was made between mound-building and subterranean termites, although it must be pointed out that many of the latter forage on the surface (Lee and Wood 1971) using the protection of covered runways. In the present context these subterranean insects are considered to be members of the surface-active fauna in deserts, and a further distinction must be drawn between them and the species that not only have subterranean nests but that also forage underground. Our knowledge of these permanent members of the desert soil fauna is fragmentary (Harris 1970), and there are two reasons for this. First, they are often

TABLE 4.2.
Numbers of Psocoptera Recovered from Litter and Mineral Soil in Various Habitats, Chihuahuan Desert, New Mexico

	Litter	Mineral Soil
High bajada		
Fallugia	33	3 (juvenile)
Prosopis	14	14
Quercus	14	4
Playa		
Prosopis	9	8
Totals	70	29

Note: Totals from five replicates.
Source: Author's personal records.

overlooked because of their completely subterranean existence; second, their colonies are usually small in size. In Nearctic deserts members of the genus *Amitermes* tend to forage in buried wood and dung (Weesner 1970), and *Coptotermes* species are similarly closely associated with the soil. The families to which these two genera belong, the Termitidae and Rhinotermitidae, respectively, are probably the most important of all the termite families in hot desert soils, and yet virtually nothing is known of the way in which their feeding activities contribute to the processes of organic decomposition here. Wood and Sands (1978) have drawn attention, however, to the fact that the subterranean galleries of these termites must affect the percolation, storage, and drainage of water and the consequent growth of plant roots in desert soils.

ARACHNIDS

This group of arthropods is predominantly carnivorous in habit and, as noted earlier, is represented in hot deserts by the relatively large-sized, surface-active scorpions, solifugids, uropygids, and spiders. But there is also a subterranean component, smaller in body size, composed of the pseudoscorpions and mites, for the most part. This fauna is virtually unknown from a taxonomic point of view, at least as far as published records are concerned, but its functional significance is beginning to be appreciated, as will be seen.

According to Cloudsley-Thompson and Chadwick (1964), pseudoscorpions are not common in hot desert soils but, rather, are restricted to marginal habitats. This may well be true, although a member of the genus *Serianus* was recorded from juniper litter in the Mojave desert (Wallwork 1970) and *Xenolpium* sp. from the Great Victorian desert of Australia (Greenslade and Greenslade, in press). Again, Ghabbour, Mikhail, and Rizk (1977) collected only three specimens of *Olpium kochi* in their survey of the mesofauna of a Mediterranean desert soil in Egypt. Pseudoscorpions have also been recovered in low numbers from litter under *Atriplex*, *Larrea*, and *Prosopis* in the Chihuahuan desert (Wallwork, in preparation). Here, they probably prey on Collembola and mites, although it is doubtful that they can play an effective part in regulating the population sizes of these arthropods in view of their own low densities (Fig. 4.6).

Increasing attention is being paid to the mite component of the desert soil fauna, and there are several reasons for this. First, mites are widely distributed in hot desert soils and often vie with the Collembola for numerical dominance. Second, this group of arachnids is a very diverse one from the ecological point of view; mites enter into the trophic structure of the desert soil community at several different levels. Some groups are carnivores, some are saprophages, others are bacterial and fungal feeders. Evidence is beginning to accumulate that indicates that at least some of these

Fig. 4.6. A desert pseudoscorpion from mesquite litter, New Mexico. Bar scale = 400 µ. (Photo by Chris Walker.)

trophic groups may be important in regulating the flow of inorganic nutrients through the plant/soil system (Gist and Crossley 1975; Crossley 1977), on the one hand, and the rate of organic decomposition (see below), on the other. Third, the composition of the mite fauna, in its broadest form, is often related to the organic content of the soil (Loots and Ryke 1967) and, at a more specific level, to its moisture status. The composition of the mite fauna therefore may be a useful indicator of soil conditions, with certain reservations as we will see.

The study of mites in hot deserts is hampered by taxonomic problems, particularly since the most widely spread and numerous order, the Prostigmata, is the least known taxonomically of all mite orders. Prostigmatids tend to prevail, particularly over the Cryptostigmata (oribatids), in mineral soils (Loots and Ryke 1967), and it is therefore not surprising that they should do so in desert sites where the organic content of the soil is low. In a survey of a range of soil types in the Chihuahuan desert, Wallwork (in preparation) found that there was a negative correlation between the relative abundance of Prostigmata and Cryptostigmata (Fig. 4.7), and this is a reflection of the preference of the former for mineral soils and the latter for organic soils. In *Atriplex* litter (organic content 28.4 percent), for example, the Prostigmata/Cryptostigmata ratio was 6.0, whereas in *Juniperus* litter (organic content 40.4 percent) this ratio fell to 0.2. Mineral soils have a lower water-holding capacity than organic soils, and it may be that prostigmatid mites can withstand xeric conditions better than Cryptostigmata, to give them a competitive advantage in such soils. Certainly the Prostigmata are more successful than the Cryptostigmata in colonizing xeric habitats at the other end of the temperature spectrum—that is, in the cold polar desert of Antarctica. In hot and cold deserts alike, prostigmatids would seem to require some physiological mechanism to survive conditions of extreme aridity; just what this mechanism might be is not known, but these are soft-bodied mites, and the assumption of some kind of anhydrobiosis cannot be ruled out (see Chapter 6).

In a Mojave desert juniper site prostigmatid mites constituted 15 percent of the soil microarthropod fauna and were mainly represented by the predatory *Spinibdella cronini* (Wallwork 1972a). Working in Australian deserts, Wood (1971) produced data that indicated that Prostigmata constituted 25 percent of soil microarthropods, although he did not give a further breakdown into trophic groups. More recent studies by Perseu Santos and Walter Whitford (1979: personal communication) and Santos, DePree, and Whitford (1978) in the Chihuahuan desert indicate that prostig-

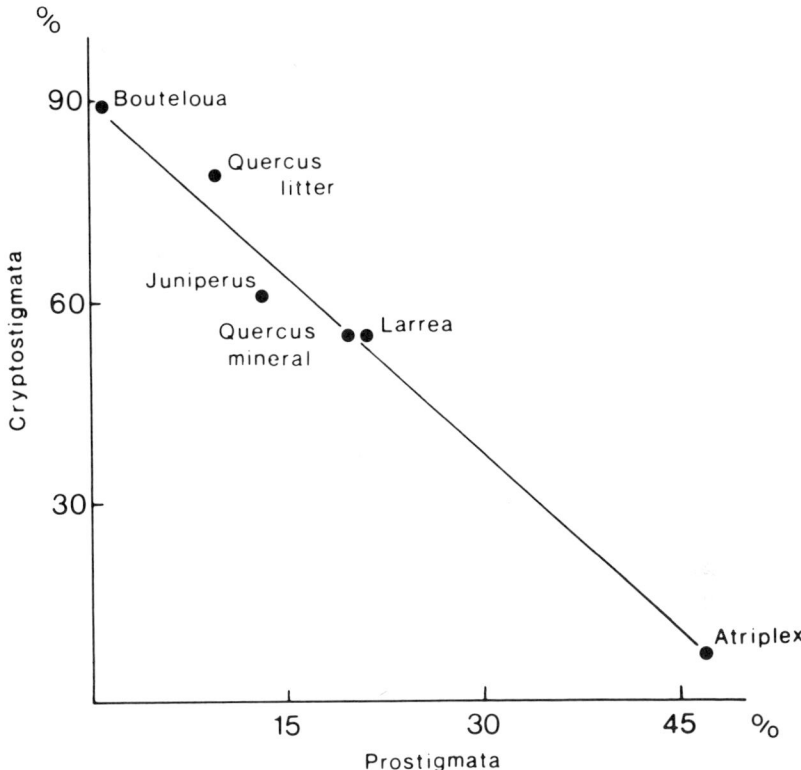

Fig. 4.7. A graphical representation of the relative abundances of cryptostigmatid and prostigmatid mites in six desert sites.

matid mites figure prominently in decomposition processes largely, it seems, through their predations on soil nematodes. These workers found that decomposition proceeded more rapidly in desert litter that was buried, and therefore in closer contact with the soil community, than litter that remained on the surface. Furthermore, it was discovered that decomposition rates were lower in litter, confined in nylon bags and treated with an acaricide, than similar material confined in untreated bags. This effect was attributed to the elimination of predatory prostigmatid mites belonging to the family Tydeidae in the treated bags. These mites feed on soil nematodes and their eggs, and when predation pressure is reduced (by the application of an acaricide), nematode populations increase in size. Nematodes, in turn, feed on bacterial populations in soils, and these, with the fungi, are the main agents in the chemical decomposition of plant litter. High nematode densities cause a

104 Desert Soil Fauna

reduction in bacterial populations and a consequent retardation of the decomposition rate.

Tydeids are not the only group of predatory prostigmatids to be encountered in desert soils. Other families represented in the Chihuahuan sites used by Santos and his co-workers include the Erythraeidae, Bdellidae, and Cunaxidae (Fig. 4.8). Alongside these predatory prostigmatids, there are others that probably function at the primary consumer level, consuming organic debris, bacteria, and fungi. These include members of the Nanorchestidae, Nematalycidae, Caeculidae, and Tarsonemidae. Some of these groups present formidable taxonomic problems, and little is known of their biology despite the fact that they may be locally abundant—for example tarsonemids in *Atriplex* litter (Wallwork, in preparation). Although caeculids rarely occur in large numbers, they are widely distributed in the deserts of the southwestern United States and have been collected from the most arid soils in the Sonoran desert (Wallwork, in preparation).

It will already have become apparent that habitats with relatively stable moisture regimes occur even within hot deserts. These are the organic soils usually located on the high desert and, as far as the microarthropods are concerned, are the sites where cryptostigmatid mites flourish best. This is a large order, and it

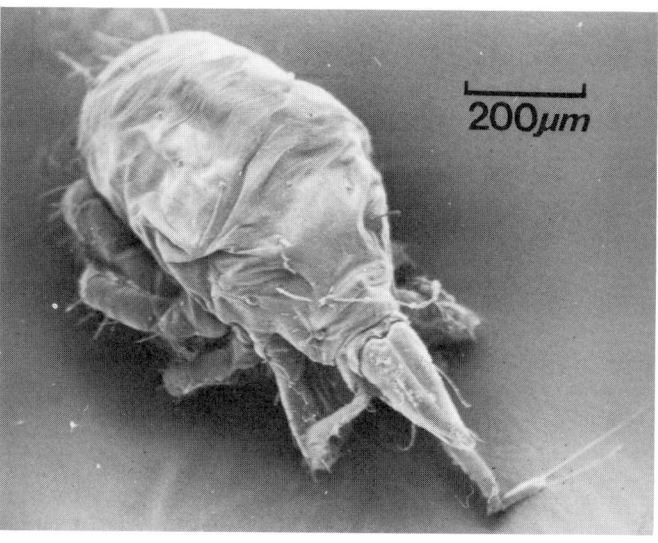

Fig. 4.8. A prostigmatid mite belonging to the family Bdellidae, collected from mesquite litter, Chihuahuan desert. Bar scale = 200 μ. (Photo by Chris Walker.)

includes groups that, according to their ecological preference, can be classified as hygrophiles, mesophiles, or xerophiles. All three of these categories are represented in hot deserts. The hygrophiles are restricted, obviously, to oasis sites where free water is readily available throughout the year. The aquatic genus *Hydrozetes* has been recorded from such sites in the Mojave desert (Wallwork 1972b). Mesophiles, represented by the small-sized Oppiidae, have been recorded in great numbers from litter under black grama grass and juniper in the Chihuahuan desert (Wallwork, in preparation). Their small body size is clearly an advantage in allowing them to seek the shelter of deeper soil horizons in the event of surface drought. Hygrophilous and mesophilous mites survive in hot deserts probably by virtue of their ability to avoid, rather than tolerate, arid conditions.

However, despite an overall tendency for cryptostigmatid mites to select relatively mesic environments within hot desert systems, there are several genera that are noted for their ability to colonize arid habitats. These include members of the Damaeoidea (Fig. 4.9), such as *Joshuella*, as well as *Passalozetes* (Fig. 4.10), *Eremaeus*, and *Zygoribatula* species, which have been recorded from sites in the Mojave desert (Wallwork 1972a, 1972b). These genera probably have a wide distribution in the deserts of the southwestern United States, for they have also been collected, often in appreciable numbers, from a range of soil habitats in the Chihuahuan desert (Wallwork, in preparation). Here they are

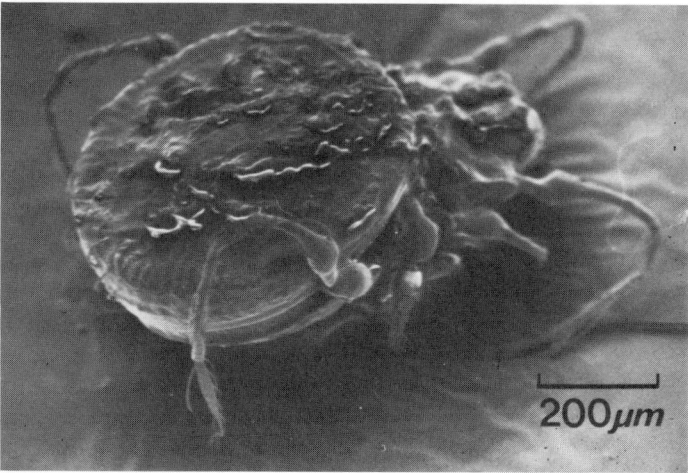

Fig. 4.9. A cryptostigmatid mite (superfamily Damaeoidea) from mesquite litter, Chihuahuan desert. Bar scale = 200 μ. (Photo by Chris Walker.)

Fig. 4.10. A cryptostigmatid mite of the genus *Passalozetes* from mesquite litter, Chihuahuan desert. Bar scale = 100 μ. (Photo by Chris Walker.)

joined by other xerophilous groups, such as *Oribatula, Scutovertex,* and *Haplozetes* (Fig. 4.11), and the eurytopic *Tectocepheus.*

Also abundant in these sites are the heavily sclerotized *Galumna* and *Scheloribates* together with the minute *Microzetes.* The small body size of the latter is a distinct disadvantage in arid environments because of the high surface area/volume ratio, but, as in the case of the oppiids, it may enable movement down the profile to avoid drought; this remains to be documented. Many of the xerophiles, on the other hand, are moderate to large in body size, and in some, notably *Joshuella, Passalozetes,* and *Tectocepheus,* the cuticle is invested with a cerotegumental envelope that may be variously folded into ridges or tubercles. Some oribatids are known to possess a waterproofing epicuticular layer of lipid (Madge 1964), and this sometimes takes the form of protuberances, or "wax blooms" (Brody 1970). Possibly the cerotegument of these xerophilous oribatids acts in this way as a device to maintain body water balance, although this remains to be established. This could be a fruitful field for research. Whatever their strategy for survival, it is clear that all of the cryptostigmatid genera mentioned above are successfully established in desert soils, for they occur here as breeding populations (Wallwork, in preparation).

Mention must also be made at this point of the occurrence in the soils of hot deserts of a group of cryptostigmatid mites that are

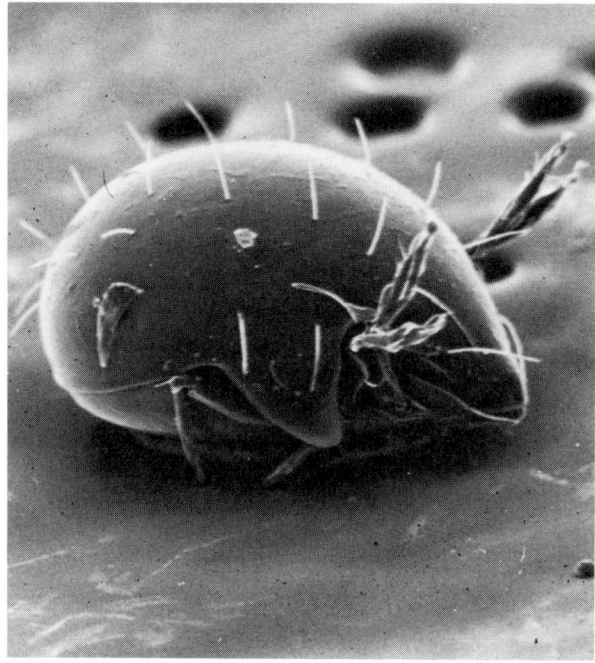

Fig. 4.11. A haplozetid mite (order Cryptostigmata) from mesquite litter, Chihuahuan desert. Bar scale = 200 μ. (Photo by Chris Walker.)

often overlooked because of their small size and delicate appearance. The group to which they belong is termed the Enarthronota, and it is conventionally placed among the more primitive Cryptostigmata, the "Oribatei Inferiores." The cuticle in the Enarthronota is not heavily sclerotized, and their powers of water retention would seem suspect. Nevertheless, they are represented in the deserts of the southwestern United States by at least three genera that have been recorded from arid environments in other parts of the world. The genus *Haplochthonius* occurs in juniper litter at high elevations in both the Mojave and Chihuahuan deserts (Wallwork 1972a, and in preparation). Grandjean (1946) also recorded it as common in human habitations and from arid soils in the southern parts of France and Spain and from North Africa. In the Mojave, *Haplochthonius variabilis* reaches peak densities in November (at the beginning of the winter rainfall period) but virtually disappears thereafter until it breeds in the following April (Fig. 4.12). In November this species occurs exclusively in the mineral soil, and it seems likely that egg laying occurs here. On hatching,

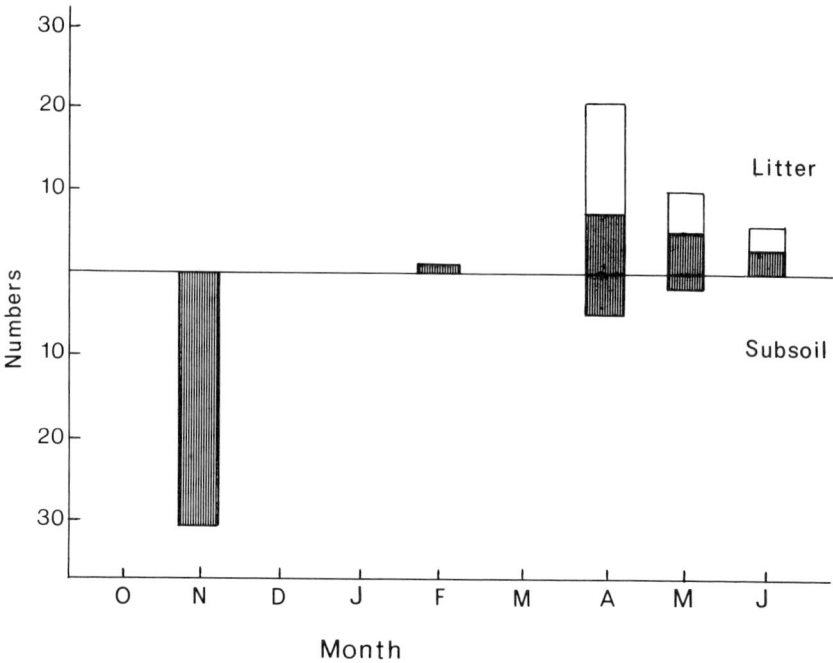

Fig. 4.12. Monthly variations in vertical distribution patterns of juvenile and adult *Haplochthonius variabilis* in a Mojave juniper site: shaded columns: adults; open columns: juveniles. (From Wallwork 1972a.)

the juveniles move up into the litter layer (Fig. 4.12), presumably to take advantage of the more abundant food supplies and higher temperatures, which will speed postembryonic development. A similar pattern of vertical migration, albeit earlier in the year, is shown by *Joshuella striata* in this same site (Fig. 4.13), and here the occurrence of gravid females only in the mineral soil lends support to the idea that such movement within the soil profile is associated with breeding activity. The virtual absence of *H. variabilis* adults in litter and mineral soil during the period December to March (Fig. 4.12) also suggests that generations do not overlap in this species. Grandjean (1946) has advanced the opinion that the related *H. simplex* from the Mediterranean region is parthenogenetic. Both of these reproductive strategies can be considered as opportunistic and a response to extreme, unpredictable environments.

The second enarthronotan genus to be encountered in desert soils is *Cosmochthonius*, and it has been recorded from juniper,

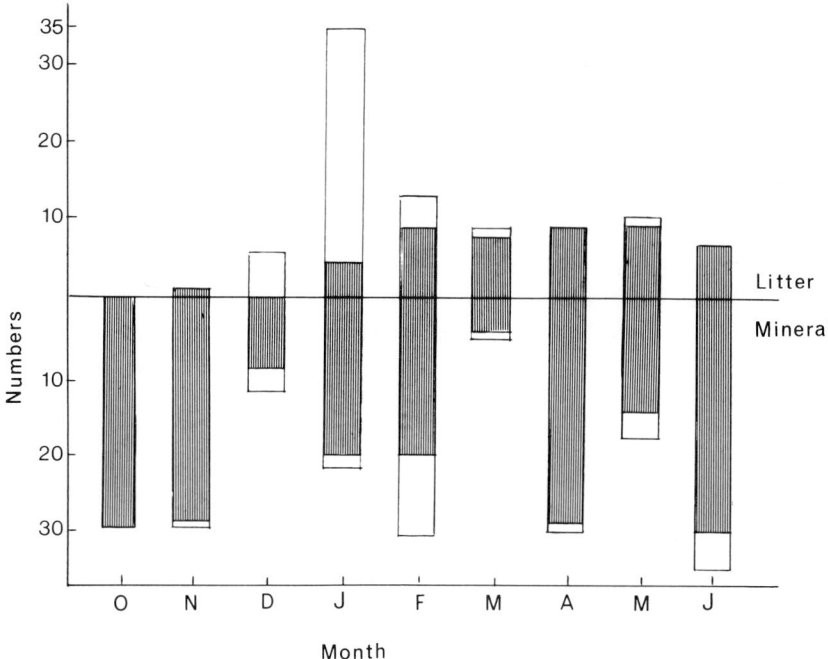

Fig. 4.13. Monthly variations in vertical distribution patterns of juvenile and adult *Joshuella striata* in a Mojave juniper site: shaded columns: adults; open columns: juveniles. (From Wallwork 1972a.)

mesquite, and saltbush litter in the Chihuahuan desert (Wallwork, in preparation). Again, Grandjean (1946) found members of this genus in buildings in France and arid sites in Algeria; it also occurs in semiarid soils in West Africa (Wallwork 1960).

Finally, members of the genus *Sphaerochthonius* occur in litter under saltbush and black grama grass in New Mexico (Wallwork, in preparation); this is also a genus that occurs in semiarid sites in West Africa (Wallwork 1960). Clearly, these primitive cryptostigmatids have solved the problem of living in arid soils, but whether they do this by avoidance or tolerance remains to be determined. The fact that species of *Haplochthonius* and *Cosmochthonius* are not uncommon in above-ground habitats (Grandjean 1946) suggests that they have some mechanism for drought resistance.

Cryptostigmatid mites feed, for the most part, on decaying plant litter, and bacteria and fungi associated with this material. Some species are known to ingest algal cells, and this habit has been recorded among desert species (Wallwork 1972a). Algal crusts occur frequently along the edges of arroyos and around the playa zone in the Chihuahuan desert, and the possibility that these microsites may serve as foci for cryptostigmatid mite populations deserves more detailed scrutiny than has so far been given. Some of the small-sized oppiids and suctobelbids are probably bacterial feeders since solid material is not usually evident in the gut contents. These mites may have an ecological role similar to that of the nematodes, particularly in sites such as black grama grass litter where these mites are very abundant (see above). In general, however, the feeding activities of cryptostigmatids are likely to have their major ecological impact in the organic soils of hot deserts, and here they are implicated mainly in the physical fragmentation of plant litter and the dispersal of microorganisms.

Two other Orders of mites occur in hot desert soils, the Astigmata and Mesostigmata, although they are poorly documented. Wallwork (1972a) collected an astigmatid species belonging to the genus *Glycyphagus* in some abundance from mineral soil under juniper in the Mojave, particularly during the December rainfall period. It occurred here mainly as the quiescent hypopial stage, which is phoretic on insects, and by this token could properly be described as a temporary member of the soil fauna. Astigmatid mites are evidently adapted to dry conditions, for they occur frequently in human dwellings and among dried storage products; so their presence in arid soils is perhaps not surprising. However, as a hypopus *Glycyphagus* sp. of the Mojave can contribute little to the functioning of the juniper soil system except perhaps serving as a source of food for predators. This is speculation. Astigmatids of the family Acaridae are locally abundant in mineral soils in the Chihuahuan desert, particularly in association with decaying root material (Wallwork and Stinnett, personal observation).

The Mesostigmata, like the Prostigmata, includes both carnivorous and detritus-feeding forms, but unlike the Prostigmata this group tends to flourish in moist, organic soils. Wallwork (1972a) recovered very few mesostigmatid mites from his Mojave juniper site, and this may be attributed to one or both of two factors. First, the aridity of the site may have exerted a limiting effect; second, the preemption of the predator niche by the prostigmatid *Spinibdella cronini* (see earlier) may have provided a competitive exclusion effect. Both of these possibilities are supported by the greater diversity and abundance of predatory Mesostigmata in the Chihuahuan desert. Santos, DePree, and Whitford (1978) found mesostigmatids in litter under *Larrea*, *Yucca*, *Fallugia*, and *Chilopsis* in this region. Wallwork (in preparation) collected these mites from litter under *Larrea, Fallugia, Juniperus, Prosopis,* and *Atriplex* in the same area (Fig. 4.14). The Chihuahuan desert experiences a greater annual rainfall than the Mojave, and its moisture regime may be more favorable to mesostigmatids than that of the Mojave. On the other hand, the Chihuahuan sites worked by Santos, Depree, and Whitford (1978) and Wallwork (in preparation) are located in a region that was formerly semiarid grassland. In such a marginal desert habitat predatory mesostigmatids could have a competitive advantage over their prostigmatid counterparts, and perhaps they have been able to carry this preemptive advantage with them as conditions became more arid. If this is the case, it serves to emphasize the fact that the composition of the soil fauna in any one place at any one time is not just a reflection of the ecological factors operating at that point in time but rather a culmination of factors that operate in a time dimension.

The majority of Mesostigmata recovered from the Chihuahuan desert sites mentioned above are predatory forms. These mites are known to feed on nematodes in other situations (Rodriguez, Wade, and Wells 1962) and, where they are locally abundant

Fig. 4.14. A mesostigmatid mite common in *Atriplex* litter in the Chihuahuan desert. Bar scale = 100 μ. (Photo by Chris Walker.)

in desert soils, may fill the ecological role of the prostigmatid Tydeidae in regulating population sizes of bacteriophagous nematodes. Their effect is probably rather localized, however, in the overall context of hot desert systems.

MAMMALS

In the previous chapter considerable attention was paid to mammals that live in desert soils but that forage on the surface. Now we can turn to the mammals that have evolved a completely subterranean, fossorial mode of life in deserts. These belong to three Orders: the rodents, the insectivores, and the marsupials.

All of the major hot and cool deserts of the world have a fossorial mammal fauna that is, for the most part, taxonomically distinct on a regional basis. Thus the subterranean mammals of the Australian deserts are the marsupial moles of the genus *Notoryctes*. In the cooler parts of the Great Palaearctic desert—that is, in Asia—the fossorial niche is occupied by the true moles, the Talpidae. Talpids also occur in North America but not in the more arid regions of the southwest. Here the fossorial niche has been claimed by the Geomyidae (Fig. 4.15). The northern parts of the Sahara and the deserts of the Middle East are the provenance of mole rats belonging to the family Spalacidae, which also extend into the cooler, arid parts of southwest Asia. Finally, mole rats of the family Bathyergidae are widespread over southern Africa, including the

Fig. 4.15. The pocket gopher *Thomomys bottae* (male) near burrow entrance. (Photo courtesy of T. A. Vaughan.)

arid regions of the southwest, while a similarly wide distribution is enjoyed by the Ctenomyidae (tuco-tucos) in South America south of the Amazon.

To exploit the subterranean environment, mammals have to possess the ability to dig burrow systems in the soil. These burrow systems provide a stable, specialized, and predictable environment, which is low in productivity and carrying capacity (Nevo 1979). It is buffered against extreme predation pressure and in the spatial dimension is discontinuous. Resources are sparse in this environment, and unequally distributed in space and time (Nevo 1979). As a consequence different groups of subterranean mammals show evolutionary convergence in three major ways: specialization, competition, and isolation.

Specialization involves convergent adaptations in structure and function. Structural adaptations are manifested by the development of digging organs, such as incisor teeth, forelimbs, and the associated pectoral girdle and claws. Concomitantly, hindlimbs, tail, eyes and ears, such characteristic features of surface-active desert mammals (particularly those that are bipedal), are reduced. The body of these subterranean burrowers is streamlined— cylindrical and tapering posteriorly; the pelage is short, and there is a tendency for hairlessness, which finds its extreme expression in the naked mole rat, *Heterocephalus glaber*, of North African deserts (Hill et al. 1957). Convergent resemblances are shown in the way in which burrows are excavated. Marsupial moles (*Notoryctes* sp.) and placental moles (*Talpa* sp.) dig with their heavily developed, strongly clawed forelimbs, as do the mole rats of southern Africa (*Bathyergus* sp.). A different method of excavation is practiced by Asian voles (*Ellobius* sp.), which use their strongly developed incisor teeth, while the pocket gophers of North and Central America share with the tuco-tucos (*Ctenomys* sp.) of South America the characteristic of using both forelimbs and incisor teeth for digging through the soil (Nevo 1979).

Competition is keen in the subterranean environment, both at the intraspecific and interspecific levels. Aggressive behavior leads to a solitary mode of life and territoriality; these features can be seen as ways of allocating a limited supply of resources (see also Chapter 8). Population densities are held at relatively low levels but, nevertheless, approximate to the carrying capacity of the environment. And there is little variation in density from season to season (French, Stoddart, and Bobek 1975).

Isolation of fossorial mammals is one outcome of aggressive competition. The establishment of territories is fixed for life (except for minor boundary changes), and those of males and females do not overlap except for periods during the breeding season. This is a feature that has been acquired independently by pocket gophers, mole rats, some tuco-tucos and moles (Nevo 1979) probably as a response to low density food resources. However, territoriality and solitariness are not immutable characteristics of subterranean mammals; colony formation may occur in areas of high food resource density. This has been recorded for the South American *Ctenomys* and the octodont *Spalacopus*, the southern African mole rat (*Cryptomys*), and the naked mole rat of North Africa (*Heterocephalus*).

Subterranean mammals can be assigned to one of two feeding categories: carnivore and herbivore. The carnivores (or, more particularly, insectivores) are the moles and marsupial moles, while the remaining groups all practice herbivory. Low resource densities in the underground environment prevent the evolution of food specialists within a particular trophic group, and the diet, both of carnivores and herbivores, consists of what is available and varies from habitat to habitat. The area from which food is obtained (home range) is the defended area and as such coincides with the territory of these subterranean mammals. In general the home range of subadults is considerably smaller than that of adults, and that of females smaller than that of males (Nevo 1979).

Fossorial mammals are active throughout the 24-hour period, and this activity is strongly linked with foraging. However, activity peaks occur during the diel cycle, and for many species these occur during the daytime. Again, as a general rule, these central-place foragers store food in their burrow systems.

Reproductive rates are relatively low in subterranean mammals. On average there are four young per litter, and most species produce only one litter per year (French, Stoddart, and Bobek 1975). This reproductive output is considerably lower than that of the most conservative of the surface-active desert rodents, the heteromyids (see Chapter 3). This relatively low level of recruitment results in stable densities operating within the carrying capacity of the environment—but close to this level since there is selection for the optimization of resources below ground. Territoriality ensures that these low density populations are optimally spaced in relation to their food resources. Optimal spacing, stabilized densities, and adjustment to available resources are the prime factors regulating the population sizes of fossorial mammals. Much less important are the effects of predation, parasites, and disease.

5
The Rainfall Factor and the Biological Response

That rainfall is a very limited commodity in hot deserts is axiomatic. As we have seen, deserts occur where rainfall does not exceed evapotranspiration on an annual basis. But when it does occur, rainfall is usually concentrated into short periods of time, measured in days rather than weeks or months, and its effects are immediate and intense. These effects are expressed in a number of ways, and these may be identified as topographical, redistributional, erosional, and biological. The first three of these are essentially physical in character and, since they are interconnected, can be considered together.

PHYSICAL EFFECTS

As we have already seen in Chapter 1, desert landscapes are very much fashioned by surface runoff in areas that receive an annual rainfall. The patterns of surface relief formed by this movement of water (variously termed erosion gullies, wadis, dry washes, or arroyos) are common to many deserts throughout the world.

The concentrated nature of the rainfall event can be illustrated by one such occurrence during August 1979 in Las Cruces, New Mexico, when a large percentage of the annual rainfall occurred in two days. Here a 10 cm rain flooded the town to a depth of 1 to 2 m, swept away roads, enlarged existing arroyos, and

116   Desert Soil Fauna

created new ones. This volume of water, descending on a desert surface, cannot be absorbed by the soil, and surface runoff effects are considerable (Fig. 5.1). These effects continually change the surface relief of desert sites, and they emphasize the instability and the dynamic nature of desert ecosystems.

Another dimension is added to this picture by the fact that rainfall events are localized not only in time but also in space. The Las Cruces rainfall event depicted in Fig. 5.1 also serves to illustrate this point. Immediately after the rain, precipitation measurements were taken at two stations, 1 km apart, on the New Mexico State University experiment site at the Jornada ranch. These two stations differed in the amount of rainfall received by approximately 5 cm. Rainfall events in deserts are unpredictable not only in their timing but also in their extent, and desert animals have to come to terms with these facts of life.

BIOLOGICAL EFFECTS

Biological communities in hot deserts are founded on a base of organic material provided, for the most part, by the dieback of aerial parts of desert plants. This dieback is largely instrumental in producing the organic soils already described in Chapter 1, but—as will have become evident—much of the organic input to desert soils does not form stable accumulations. It becomes dispersed and

Fig. 5.1. Surface runoff after torrential summer rains in Las Cruces, New Mexico, in August 1979.

mixed to varying degrees with mineral soil. The deserts of the southwestern part of the United States, where much of the research into the ecology of arid zones has been carried out, are dominated by the creosote bush (Fig. 5.2). This shrub occurs, virtually as a monoculture, across 1,000 miles of lowland desert extending from southern New Mexico, through southern Arizona, to the southeastern parts of the Californian Mojave. This shrub, which can live for more than 100 years (Fonteyn and Mahall 1979), usually does not attain a height of over 2 m, and its canopy is not dense or extensive enough to provide enough shelter for litter to accumulate beneath it in any appreciable quantity. Creosote bush litter is easily carried away by surface runoff and is redeposited as a "strand" line along the edges of arroyos (Fig. 5.3). Here it becomes matted together as it dries out and is colonized by various members of the soil fauna that promote its decomposition (see Chapter 4). Santos, DePree, and Whitford (1978), for example, established a linear relationship between the amount of creosote bush litter accumulation and the population sizes of its microarthropod constituents. The distribution patterns of leaf litter and the rate of its accumulation are governed to a large extent by the local hydrology. This must figure

Fig. 5.2. The creosote bush, *Larrea tridentata*, in the Chihuahuan desert.

Fig. 5.3. Arroyo margin showing litter accumulation along strand line. Sonoran desert near Glamis, California.

prominently in ecological studies of soil communities in hot deserts. Recognition must also be given to the fact that physical factors associated with rainfall can cause the burial of organic material and that decomposition processes will be accelerated as a result.

THE BIOLOGICAL RESPONSE

From the biological point of view, rainfall is the driving variable in hot deserts. Biological activity is stimulated in various ways by rainfall events. The flowering of many desert plants, such as creosote bush, and the appearance of crucifers, lotus, desert marigolds, verbena, and zinnias occurs after substantial rains. Animals, such as millipedes, ants, beetles, snails, and rodents, become surface active after rains, and subterranean microarthropods reproduce during this season (see later in this chapter). There is here a clear biological response to the rainfall factor, a response that is essentially an avoidance tactic designed to minimize exposure to drought conditions at other times of the year. There is yet another side to this coin since many desert animals have developed ways of tolerating drought conditions. These twin virtues of drought avoidance and tolerance lend themselves to separate examination, although, as will become apparent, the two are frequently incorporated in a particular survival strategy. It is convenient, however, for our purposes to treat them separately. The mechanisms of drought tolerance are dealt with in the next chapter; those of drought avoidance in this. But before we can examine this response in detail, we must focus first on the more general features of the rainfall effect.

If rainfall is a driving variable in hot deserts, it also follows that it is a limiting variable, and in this respect, hot desert systems differ from their cool, moist temperate counterparts. In the latter the availability of solar energy limits biological activity because environmental moisture rarely reaches critically low levels. In hot deserts there is an abundance of solar energy throughout the year, but, despite this, biological activity can be discontinuous. In other words, "no-growth" situations occur in hot deserts, even when solar energy is plentiful, because moisture levels are too low to allow growth and reproduction to proceed in a normal manner.

The Pulse-and-Reserve Paradigm

The distinction between hot desert and cool, moist temperate ecosystems has been succinctly defined by Noy-Meir (1973) as a contrast between two paradigms (submodels or modules). The hot desert system can be identified with a "trigger-pulse-reserve" phenomenon, while cool, moist temperate ecosystems conform to a "level-controlling-flows" paradigm.

The level-controlling-flows module is based on the premise that the energy level in any compartment of an ecosystem is regulated by rates of inflow and outflow. A simple analogy is that of a bath of water with inlet and outlet pipes. The level of water in the bath is a function of the amount entering and the amount leaving at a particular time (Fig. 5.4). In this type of system the flow of energy into and out of a compartment is continuous.

The pulse-and-reserve paradigm postulates a discontinuous flow of energy through compartments of an ecosystem. According to this module, an environmental trigger (such as rainfall) initiates a pulse of biological activity that manifests itself as primary, secondary, or tertiary production. Much of this production is lost from the system owing to various mortality factors, but some is directed into a reserve—seeds or eggs, for example. This reserve is essentially a no-growth compartment from which the next pulse of activity originates. The magnitude of this pulse will be determined by the effect of the trigger on the reserve compartment and, indeed, the level of the reserve that is activated. Of course no compartment of an ecosystem (trophic level) is self-perpetuating. It requires the input of energy from another compartment, and in hot deserts the source is the pulse or reserve from that compartment. The magnitude of this input will regulate the extent of the response (pulse) of the recipient compartment. This concept of the processes occurring in hot deserts can be summarized in an information flow diagram module (Fig. 5.5).

Time Scales

Rainfall events in hot deserts are discontinuous to a much greater degree than they are in moist temperate localities. As we have seen, the total yearly rainfall may occur within a restricted period of a few days in the southwestern deserts of the United States; parts of the Sahara may not experience a rainfall event for several years. Moreover, it has been stated by Noy-Meir (1973), who

Temperate

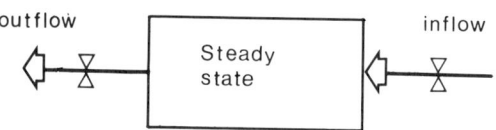

Fig. 5.4. The "level-controlling-flows" paradigm. (After Noy-Meir 1973.)

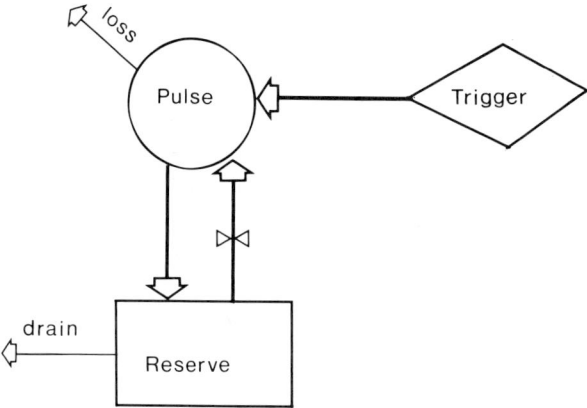

Fig. 5.5. The "pulse-and-reserve" paradigm. (After Noy-Meir 1973.)

discusses rainfall patterns in hot deserts at some length, that these events are largely stochastic—that is, their magnitude and timing have an appreciable random component. If biological activity is triggered by rainfall events, as suggested above, it follows that it also will be stochastic in nature. But, as Noy-Meir (1973) further points out, there are variations on this stochastic theme. These variations occur as a result of the degree of clustering of rainfall events and the recovery time required by biological systems exposed to these rainfall events. In deserts where rainfall is irregular and aseasonal, pulses of biological activity are discrete and strongly correlated with the amount of rainfall. In cool, winter deserts, on the other hand, rainfall events are more strongly clustered and, coupled with the fact that biological responses to these triggers are slower, a situation is created in which peaks of biological activity merge together to form a single large pulse. This is the case in deserts where rainfall is seasonal, and this total rainfall constitutes a single trigger. These aseasonal and seasonal effects are depicted in Fig. 5.6.

What guidelines (it would be presumptious to suggest "conclusions") can be derived from these models of hot desert ecosystem functioning, particularly in relation to the rainfall factor? Immediately, the time factor must be recognized, creating seasonal deserts in some regions, aseasonal ones in others. But seasonality is subject to variation, as evidenced by rainfall patterns in the

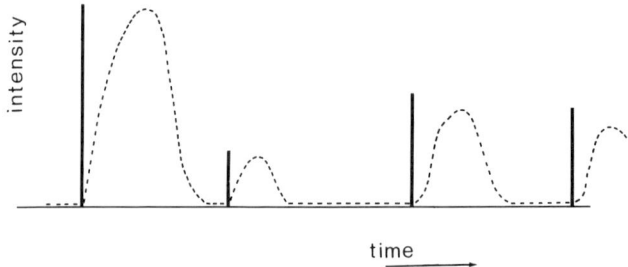

Fig. 5.6. Rainfall patterns and the biological response in hot deserts. (After Noy-Meir 1973.)

southwestern parts of the United States. Here long-term weather records indicate that there are consistent patterns of rainfall within a desert type but that one desert differs from another in this respect. For example, moving eastward from the Mojave of southern California across the Sonoran desert of Arizona to the Chihuahuan desert of New Mexico, there is a rainfall gradient that shows a steady increase in mean annual precipitation. Moreover, in its southern part at least, the Mojave experiences winter rains, carried in by an airstream that originates over the north Pacific. In contrast the Chihuahuan desert is mainly influenced by summer rainfall carried in from the Gulf of Mexico. The Sonoran desert receives rain from both sources and, consequently, has winter and summer rainfall peaks, albeit attenuated ones (Fig. 5.7).

It is reasonable to suggest, therefore, that the driving variable (rainfall) in hot desert systems is, in the short term, stochastic in nature. It cannot be accepted as axiomatic, however, that this reasoning will necessarily hold up in the long term. It is more likely that hot desert systems exhibit a short-term environmental unpredictability on which is superimposed a long-term stability. The nature of the biological response must be examined in these terms, and this implies a paradox. Desert soil animals live in environments that are, predictably, unpredictable. They respond to the short-term unpredictability by being opportunistic. They respond to the long-term predictability by being avoiders or tolerators or both. These lifestyles can now be examined in more detail. The remainder of this chapter is devoted to a consideration of mechanisms developed to avoid the kind of environmental stress occasioned by periods of drought. The following chapter is devoted to mechanisms designed to tolerate this type of stress.

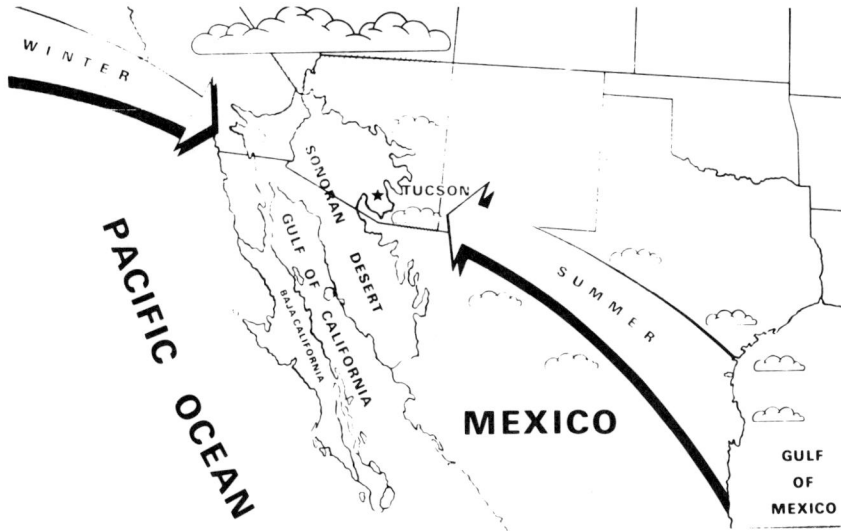

Fig. 5.7. Patterns of summer and winter rainfall in the deserts of the southwestern United States.

DROUGHT AVOIDANCE

The avoidance mechanisms employed by members of the desert soil fauna that enable them to survive in an environment that, in general terms, is subject to moisture stress for considerable periods of the year are varied. They fall into three categories: reproductive, behavioral, and physiological.

The Reproductive Response

The timing of breeding activity to coincide with the rainy season in hot deserts is a general phenomenon, not only among animals but also among plants, and its survival value is obvious. Juvenile stages, which require food and moisture for rapid growth, are recruited when environmental stress is minimal. For example, the desert woodlouse, *Hemilepistus reaumuri*, breeds and produces a new generation during the late winter, spring, and early summer in the Negev; arid zone millipedes, similarly, breed during the wet season (Chapter 2). This phenomenon is also well illustrated by the results of a study on the population dynamics of soil microarthropods in a soil under *Juniperus* in the Mojave desert (Wallwork 1972a, 1980).

One of the main results of this study was that population sizes doubled immediately after December rains (Fig. 5.8). This was due in part to the influx of an astigmatid mite, *Glycyphagus* sp., which has a phoretic hypopus, but more important perhaps to the recruitment of a new generation by the cryptostigmatid mite, *Joshuella striata*, and the predatory prostigmatid, *Spinibdella cronini* (Fig. 4.13). These findings suggest that such soil microarthropods in hot deserts can gear their life cycles in such a way that the most susceptible juvenile stages appear at the most favorable times of the year. The mechanism whereby this is achieved is a rapid response to the rainfall trigger. It will also be apparent—from what has been said in Chapter 2—that those members of the desert soil fauna that breed mainly in the rainy season (*Hemilepistus reaumuri* and millipedes have already mentioned in this context, and the list also includes many soil-dwelling rodents) assume a dormant state during the dry season. However, this may not be the case with some of the more completely subterranean microarthropods. Cryptostigmatid mites may be active throughout the year, but even so, their breeding seasons may be restricted to the rainy months, as is the case with *Joshuella striata*. Extensive and

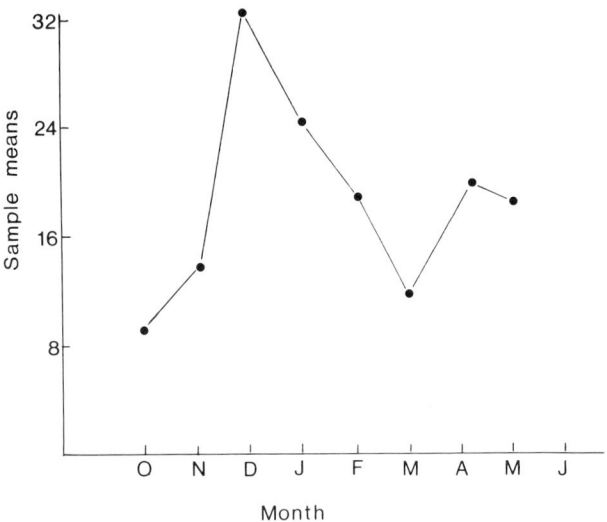

Fig. 5.8. Monthly variations in total microarthropod numbers in a Mojave juniper site. (From Wallwork 1972a.)

intensive sampling over a yearly cycle (or even longer) is required to confirm that this is a general phenomenon. However, some progress in this direction is being made, and preliminary studies in a southern New Mexico bajada site, conducted immediately after torrential summer rains, revealed evidence of considerable breeding activity. Adult and larval pauropods were collected simultaneously in large numbers from mesquite litter. Most of the female cryptostigmatid mites from mesquite, juniper, desert oak, and saltbush litter were gravid, while the predatory Mesostigmata, such as the species illustrated in Fig. 4.14, were also producing a new generation, to judge from the number of juveniles recovered. This response to the rainfall trigger receives further confirmation from laboratory observations carried out by Penny Greenslade (1979: personal communication) on an arid zone collembolan belonging to the genus *Sphaeridia*. She noted that this species develops from egg to egg within the space of one week and that hatching of the egg is dependent, to some extent at least, on exposure to alternating dry and wet conditions. These observations suggest that the biological response to the rainfall trigger is an opportunistic one. They also suggest that a pattern is beginning to emerge that may have general applicability to desert soil systems. However, a note of caution must be sounded here. Opportunistic reproductive strategies, intuitively, imply maximum recruitment to a population at the most favorable time of the year. Yet as we saw earlier (Chapter 3), the surface-active rodents in hot deserts are opportunistic in the sense that breeding activity reaches a peak at this time, but they exhibit a more controlled response by reducing reproductive output, compared with their mesic relatives. This controlled response can be viewed as a tolerance mechanism, and it serves to remind us that ultimate survival in hot deserts is not simply a question of avoidance or tolerance but a combination of these two strategies. Yet another facet of survival in hot deserts is provided by the termites, which by virtue of an elaborate social structure can create and control their immediate environment and thereby achieve a considerable degree of independence from the exigencies of a pulse environment.

    The reproductive strategies considered above, which reduce the exposure of susceptible developmental stages to drought, can be considered as numerical responses. They are mechanisms that ensure the continuity of the population through an annual (or greater) cycle of favorable and unfavorable environmental conditions. The remaining two categories of avoidance mechanisms can, more properly, be considered as functional responses in which a premium is placed on the survival of the individual.

The Behavioral Response

Burrowing, nocturnalism, and seasonality are all examples of behavioral response that allow animals to avoid water stress in hot deserts. These topics have already been introduced, in their appropriate places, in previous chapters, but it is pertinent to attempt a synthesis at this point.

The ability to burrow in the soil is an important feature of the life-style of many desert animals. It is a factor that is vital to survival in many cases, but it is expressed in various ways and to different degrees. At one extreme are the scolopendromorph centipedes and some scorpions, such as the North American *Centruroides sculpturatus* and the African *Buthus hottentotta*, which do not actively excavate soil burrows but rather seek the shelter of existing burrows and crevices under rocks. Other desert scorpions, such as *Diplocentrus spitzeri* of North America, produce shallow burrows, while *Buthus occitanus* and *Scorpio maurus* of North Africa are deep burrowers. Millipedes of arid and semiarid regions are also efficient burrowers, and these include *Orthoporus ornatus* of North America and the gomphiodesmid and paradoxosomatid millipedes of West Africa. Again, woodlice of the genus *Hemilepistus* depend for their survival in hot deserts on being able to construct burrows to a depth of 40 cm where the amount of moisture in the soil remains at tolerable levels (Shachak 1980) (see also Fig. 2.11). The behavioral parallels that can be drawn between taxonomically distinct and geographically remote groups of desert arthropods are indicative of the limited number of avoidance strategies available for exploitation. Even so, these parallels are not complete. For example, *Orthoporus ornatus* is not an obligate burrower; it can seek refuge in *Novomessor* (harvester ant) nests when these are available (Wooten and Crawford 1974). The gomphiodesmid and paradoxosomatid millipedes, on the other hand, not only burrow in the soil but construct spherical molting chambers in the burrows they have excavated. This implies a greater reliance on the ability to burrow than is the case with *O. ornatus*. Again, the complex burrow system of *Hemilepistus reaumuri*, with its vertical and horizontal arrangement of tunnels and rooms, contrasts with the shallow, temporary excavations produced by such conglobating desert wood lice as *Venezillo arizonicus* and *Armadillo albomarginatus*.

Burrowing ability and the maintenance of burrow systems is realized, par excellence, by ants and termites. The greater degree of permanence of these burrow systems, compared with those of millipedes and scorpions, is probably a reflection of the fact that colonies of these social insects persist for long periods of time. Other insects that possess the ability to burrow in desert soils include the tenebrionid beetles and the sand roaches *Arenivaga* and *Heterogamia*, although some of these at least may practice the habit of "sand swimming." A more typical fossorial habit is shown by desert rodents; indeed, some of these are almost completely subterranean, for example, the Geomyidae. Well-developed burrow systems are also produced by kangaroo rats and pocket mice, which forage on the soil surface. By contrast, pack rats of the genus *Neotoma* build surface nests, often in shaded depressions under rocks or large shrubs. *Neotoma* nests are readily identified by the piles of sticks or pieces of cholla cactus heaped upon them (Fig. 5.9).

To burrow in the soil, an animal has to possess efficient digging organs, and these are basically of two kinds: legs and mouthparts. Scorpions, tenebrionid beetles, desert roaches, and most rodents use their walking legs to excavate burrow systems, as do the desert tortoises and spadefoot toads. The limbs of these animals are robust and, in the case of the roaches, tortoises, and moles, spatulate to increase their effective surface area and are equipped with strong spines or claws. This convergence in functional morphology extends to the spadefoot toads of North and South America (*Scaphiopus* and *Ceratophrys* spp, respectively), which use the horny spur on each hindlimb as a digging organ. On the other hand, soil particles are removed from the burrows of ants

Fig. 5.9. Nest of the pack rat *Neotoma* in the Sonoran desert near Glamis, California.

and termites with the aid of the mandibles, and certain rodents show convergences with this type of digging behavior. Asian voles (*Ellobius*), the mole rats of the Middle East (*Spalax*), the gophers of North America (*Geomys*), and the tuco-tucos (*Ctenomys*) of South America use forelimbs and incisor teeth as excavators. Here again, there are parallelisms, or convergences, between taxonomically and geographically distinct groups of soil animals with respect to burrowing behavior.

The ultimate result of burrowing activity is to locate moist environments that lie beneath the soil surface. In this respect the distribution patterns of these environments in hot deserts need to be examined. Unlike cool, moist temperate soils, where the moisture status remains fairly uniform in the horizontal dimension, hot desert soils accumulate moisture in discrete areas. The leaf catchment zone beneath desert shrubs is one such example (Cloudsley-Thompson and Chadwick 1964), and this zone may be exploited by burrowing rodents such as the kangaroo rat *Dipodomys merriami*. This rodent has a number of adaptations that ensure survival in arid environments (see Chapter 6), but it places some reliance on receiving some of its moisture requirements from above-ground sources, that is, seeds. Termites may also obtain moisture from the food they consume on the soil surface, such as woody tissue and dung. They are also able to tap subterranean moisture; they do this not necessarily from leaf catchment zones but from the deep water table that they reach by vertical shafts constructed to a depth of several meters in some cases. Water is carried in the crop from these deep levels and regurgitated. Among the genera that tap environmental moisture in this way are the Saharan *Psammotermes* and *Macrotermes* and the Asian *Anacanthotermes* (Grassé and Noirot 1948; Ghilarov 1962). The emphasis that is placed on water gain from the environment, in contrast to water conservation, by these termites identifies them primarily as avoiders of arid conditions. These two components of water balance are much more delicately poised in the case of the kangaroo rats, where nocturnalism assumes a special significance. This line of thinking leads, quite naturally, into a consideration of nocturnal behavior. But before we embark on this, one further point remains to be made about burrowing activity.

Soil burrows represent relatively closed systems in that the exchange of air between them and above-ground atmosphere is limited. This creates a problem for an animal, such as a kangaroo rat, which may spend considerable periods of time underground. Its prolonged presence here can result in the buildup of levels of CO_2, accentuated by soil moisture and temperature conditions, to higher than ambient. This possibility was noted by Kay and Whitford (1978) who suggested that surface winds, such a constant feature in hot deserts, are probably important in the ventilation of burrows.

There are so many references in the literature to the nocturnalism of desert animals that one might be forgiven for assuming that this pattern of activity is the rule rather than the exception particularly among those groups that live in the soil but forage periodically on its surface. To be sure, many surface-active predatory invertebrates are nocturnal, and here we include the centipedes, scorpions, solifugids, and uropygids. Nocturnalism is also a characteristic feature of desert wood lice and rodents, although here there are exceptions. The isopod *Hemilepistus reaumuri* has a crepuscular rhythm and shows a positive photoreaction, except at high temperatures. The desert gerbil, *Meriones hurrianae*, some pocket mice, and ground squirrels are day-active on the ground surface.

These exceptions are not isolated examples. Desert millipedes, snails, termites, tortoises, and lizards are habitually day-active rather than nocturnal. These groups have other ways of avoiding or tolerating the unfavorable conditions, as we will see later. Ants also show considerable surface activity during the daytime, and this is particularly true of species that forage individually; group foragers are often nocturnal. Again, tenebrionid beetles exhibit a variety of activity patterns, some species being day-active, some nocturnal, and still other crepuscular (see Chapter 8).

Many of these desert animals that are active on the soil surface during the daytime restrict this activity to the more favorable periods of the day: early forenoon and late afternoon. They also limit surface activity to the most favorable times of the year, that is, the rainy season; at other times they may become dormant in soil burrows (millipedes and rodents) or on the soil surface (snails).

The Physiological Response

The principal physiological mechanism that allows a desert soil animal to avoid drought conditions is the assumption of a dormant, quiescent state often referred to as diapause. This state takes a variety of forms, depending on the animal group under consideration and the environmental problem that has to be solved. Dormancy, in the conventional sense, applies both to the state of hibernation and that of estivation. The former relates to the assumption of an inactive condition during periods of low temperature; the latter to inactivity during periods of high temperature. Hibernation is usually associated with homiotherms, and in the desert soil environment these are mammals. Estivation, on the other hand, is a condition that can be assumed by homiothermic and poikilothermic animals. The biological response to the temperature factor is considered in more detail in Chapter 7. For the moment it is sufficient to point out that at extremes of temperature in hot deserts moisture is at a minimal level or unavailable. In passing, it is perhaps worth noting that the terms *hibernation* and *estivation* are often used interchangeably and indiscriminately in the literature.

Dormancy is not a feature of the life-styles of surface-active predatory invertebrates, but it is of great importance to the survival of amphibians, some insectivorous marsupials, and such detritivores and herbivores as millipedes, snails, tenebrionid beetles, ants, and many rodents. Dormancy in these cases is associated with a withdrawal into soil refugia, and this act in itself constitutes an avoidance reaction. Desert snails, on the other hand, do not burrow extensively. They estivate attached to vegetation or rock surfaces or just below the soil surface (Riddle 1975; Shachak, Orr, and Steinberger 1975), and this dormancy allows them to tolerate environmental conditions that they do not seek to avoid, at least in the case of *Rabdotus schiedeanus* of North America and *Sphincterochila zonata* and *S. prophetarum* of the Middle East.

It is pertinent at this point to define what we mean by *avoidance* and *tolerance* with regard to survival strategies of desert soil animals since the distinction between these two life-styles is not always clearly drawn. Desert animals that rely primarily on avoidance strategies remove themselves from situations of potential stress. In contrast tolerators have mechanisms for coping with and, thereby, remaining in stress situations. Using this criterion, it is clear that desert snails, such as *Rabdotus schiedeanus* and *Sphincterochila* spp., exhibit drought tolerance rather than avoidance. By the same token, some ground-based tenebrionid beetles and rodents are tolerators rather than avoiders. Two examples serve to illustrate this.

The tenebrionids of the Namib dune desert show various patterns of circadian and seasonal activity, which identify them either as drought avoiders or drought tolerators. To the former category belong species such as *Lepidochora porti* and *L. argentogrisea* that although active throughout the year are nocturnal. In these cases the avoidance mechanism is behavioral rather than physiological. Drought tolerators include species that are day active, such as *Onymacris plana, O. rugatipennis, Gyrosis moral-*

esi, and *Calosis amabilis*. They achieve this tolerance through their ability to conserve body water (see Chapter 6), and this ability allows *O. plana* and *O. rugatipennis* to remain active all the year round. *G. moralesi* and *C. amabilis*, on the other hand, are inactive during the winter months, and this may be a response to low temperatures at this time of year.

Although most desert rodents are nocturnal, thereby avoiding water stress, attention has already been drawn to the fact that ground squirrels, some pocket mice, and the gerbil *Meriones hurrianae* are day-active (Ghobrial and Nour 1975; Prakash 1975). Pocket mice (*Perognathus* spp.) tolerate the warmer months of the year but assume a dormant condition during the colder months. Hibernation also occurs in the North American ground squirrels *Citellus armatus*, *C. spilosoma*, and *C. townsendi*, and this can be interpreted as a response to low environmental temperatures. However, *Citellus leucurus* and *C. tereticaudus* do not hibernate; they are tolerators, and the key to their survival in hot deserts is very much a matter of temperature tolerance (see Chapter 7 and Hudson, Deavers, and Bradley 1972).

These examples illustrate the considerable overlap between the responses of desert soil animals to environmental moisture and temperature. They also demonstrate that strategies of avoidance and tolerance are not mutually exclusive. The tenebrionid beetles *Gyrosis moralesi* and *Calosis amabilis* can tolerate summer conditions but avoid winter stress by becoming dormant.

The violent fluctuations in moisture and temperature, characteristic of hot deserts, can never be avoided completely. Some degree of tolerance must be developed, even among the avoiders. We can now turn our attention to tolerance strategies.

6
Drought Tolerance

Animals that successfully tolerate the aridity of hot deserts employ a number of strategies that fall, basically, into two types: those that maximize moisture gain from the environment and those that minimize water loss to the environment. Survival may well depend upon a combination of strategies drawn from both of these categories. We look first at the general nature of these strategies before discussing the ways in which different groups of desert soil animals have exploited them.

WATER GAIN AND WATER CONSERVATION

Water is present in the desert environment, like everywhere else, in exogenous and endogenous forms. Exogenous, or free, water occurs in five compartments of the desert model, and these may be separated temporally and/or spatially. First, there are the permanent bodies of water found on the surface at oases. These represent important sources of water for highly mobile desert animals but are highly localized in the spatial dimension. Second, there is deep groundwater. This is available all the year round to animals that can burrow deep enough to tap it. Third, there is free surface water that floods the desert and accumulates temporarily in playa lakes. This source of water is available only during the restricted period of the summer rains, and it may be too saline to be of much use. Fourth, there is preformed water, available to herbivores that feed on succulent vegetation and to carnivores that suck the juices of their prey. This source of water may be in limited supply for carnivores, depending on the seasonality of their prey; in such limiting circumstances these carnivores may switch to a vegetarian diet, and examples of lizards and rodents that do this have already been cited in Chapter 3. No such problem of availability faces the more committed herbivore that feeds on the sap of cacti and succulents, but it may have to face others (see below). Finally, there is atmospheric water vapor, which is mainly available at night when it condenses on the aerial parts of vegetation. Dew may be an important source of water for various desert soil animals (see below), and, indeed, there are a few species that can extract this moisture from unsaturated atmospheres, as we will see.

Endogenous water is chemically bound and is available to those herbivores or, more particularly, granivores that can release it through oxidative metabolism. Its availability depends on the seasonality of seed production, although many granivores can switch to preformed water supplies when they are available during the rainy season (see below).

It is the ability to conserve body water, rather than to regain what it loses to the environment, which signifies a well-adapted desert animal. Conservation mechanisms can operate in one or more of four ways that are identified with the four main pathways of water loss from the body. These avenues of water loss are essentially the same for desert soil invertebrates and vertebrates, and the mechanisms for water conservation show remarkable parallels in these two faunal groupings.

In the first instance water is lost across the general body surface by transpiration. This loss can be curtailed in a number of ways. For example, a relatively large body size reduces the surface area/volume ratio and thereby the relative evaporative surface. Perhaps more important, transpiration rates are lower in desert soil animals than in their cool temperate relatives, and this may be a function of the presence of a waterproofing layer in the integument, the absence of sweat glands, or the development of some kind of carapace. Second, water is lost from respiratory surfaces, which may be internal (tracheae and lungs, for example,) or, more occasionally, external (the skin of amphibians and the pleopods of wood lice). Water loss from internal surfaces is restricted by ventilation control mechanisms; from external surfaces control is effected by behavioral adaptations coupled with a reduction in the use of the general body surface as a respiratory organ. Water is also lost during excretion and secretion. Excretory water loss is controlled by renal structures that produce insoluble waste products or an extremely concentrated urine. Various desert soil animals also secrete fluids of a repugnatorial (millipedes, tenebrionids) or venomous nature (arachnids). These pathways of water loss, again, are controlled by the production of concentrated fluids. Lastly, water loss from the alimentary canal is reduced to a minimum by the production of feces with a very low moisture content.

These then are the tactics adopted for drought tolerance by desert soil animals. We can now consider to what extent they are utilized by different groups, beginning with the surface-active fauna.

SURFACE PREDATORS

Under this heading we must consider the relatively large-sized carnivorous arthropods that are based in the soil but forage for their prey on its surface. These are the centipedes and the arachnid scorpions, solifugids, uropygids, and spiders. Also included in this section are such carnivorous vertebrates as the amphibians, lizards, and marsupials. These groups can conveniently be considered together because they illustrate a range of adaptability to hot desert conditions as far as their water relations are concerned.

Most of these groups (the lizards are the only main exception) have two features in common: they are nocturnal and their water source is mainly the body fluids of their prey. Beyond this, there are few similarities.

Arthropods

Centipedes become desiccated quite rapidly in dry air—water loss occurs mainly by respiration and transpiration. Cloudsley-Thompson (1959) studied transpiration rates of three tropical or subtropical scolopendromorphs: *Scolopendra clavipes* from Tunisia, *Rhysida nuda togoensis* from Ghana, and *Ethmostigmus trigonopodus* from Malawi. He demonstrated that the rate of water loss from *R. nuda togoensis* was comparable with that of *Lithobius* spp. from more temperate climes; this is not entirely unexpected since the habitat of *R. nuda togoensis* can be considered as moist tropical. However, rates for *E. trigonopodus* and the desertic *S. clavipes* were about 50 percent lower. Even lower rates were shown by a *Scolopendra polymorpha* species complex from New Mexico (Cloudsley-Thompson and Crawford 1970). Scolopendromorphs are more primitive than lithobiomorphs, and the greater ability of the former to conserve water may be regarded as a specialization for life in a hot, dry climate. Cloudsley-Thompson and Crawford (1970) calculated that *S. polymorpha* lost water from the body at the rate of 1.08 mg cm^{-2} hr^{-1} at 30°C (the corresponding value for *Lithobius* species is 6.8 mg cm^{-2} hr^{-1}), but this is still high compared, for example, with that of insects (Lewis 1972). Rates of water loss from Saharan ants do not exceed 0.36 mg cm^{-2} hr^{-1}, according to Délye (1969). Furthermore, there is no critical or transition temperature at which the scolopendromorph cuticle becomes permeable. This strongly implies that there is no epicuticular waterproofing lipid layer. Nor can these desert arthropods take up moisture through the integument from unsaturated air or moist surfaces (Cloudsley-Thompson and Crawford 1970) as can ticks, the sand roach *Arenivaga*, and the desert thysanuran *Ctenolepisma terebrans* (see later). Desert scolopendromorphs must obtain all the water they require from the tissue fluids and blood of their prey and by drinking free water when this is available. They are not well adapted to the most arid conditions, and their distribution patterns reflect this. They occur for the most part around the fringes of hot deserts where moist refugia can be found.

Conflicting reports are available concerning the water relations of the whip scorpion *Mastigoproctus giganteus*. This uropygid, the vinegaroon of the deserts of the southwestern United States and northern Mexico, has been the object of the attentions of Ahearn (1970b) and Crawford and Cloudsley-Thompson (1971). This arachnid obtains water from the fluids of its prey, but it will also drink free water and possibly absorb this through its cuticle, according to Ahearn (1970b). This author presented evidence to indicate that small individuals (up to 4 g) had a transition temperature of 37.5°C (Fig. 6.1), indicating the presence of an epicuticular lipid water barrier. Crawford and Cloudsley-Thompson (1971), on the other hand, could obtain no evidence for the presence of such a waterproofing layer in large (4.1 g to 7.5 g) animals. They also found that transpiration rates remained relatively constant and high over a range of relative humidities (around 2 percent of the original body weight per hour at 26°C), whereas Ahearn described an inverse relationship between transpiration rate and relative humidity. These comparisons are not entirely valid in view of the fact that different size categories (and probably different stages in maturity) were tested in the two cases. Furthermore, the experimental conditions were different; Crawford and Cloudsley-Thompson maintained their experimental animals on moist sand—an abrasive substrate that could remove an epicuticular lipid layer. Ahearn conducted his experiments on moist tissue paper, which does not

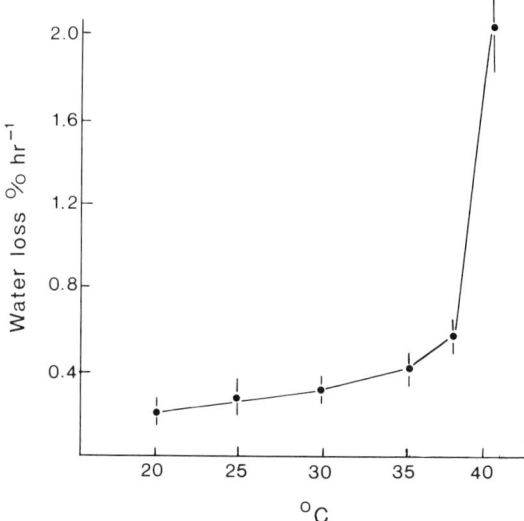

Fig. 6.1. Water loss in *Mastigoproctus giganteus* as a function of temperature, based on a minimum of nine individuals exposed for two hours at 0 percent relative humidity in slowly moving air. (From Ahearn 1970b).

have such abrasive properties. *Mastigoproctus giganteus* is exposed in nature to the abrasive effects of a sandy substratum, and it is difficult to conceive how it could have evolved an epicuticular water barrier that would be quickly removed once it started to burrow in the soil. On the other hand, as we will see later in this chapter, another sand-burrowing arthropod, the cockroach *Arenivaga*, can repair an abraded cuticle in a matter of one or two days; possibly *Mastigoproctus* has this ability? Again, there are certain scorpions that burrow in sandy desert soil and apparently do not suffer from abrasive effects.

Be this as it may, there is no doubt that *M. giganteus* loses water in dry atmospheres at a greater rate than xeric arthropods of comparable size (Table 6.1). Ahearn (1970b) has shown that the bulk of this water loss occurs across the cuticle rather than through respiration, and there are parallels here with desert tenebrionid beetles (see later). The vinegaroon tends to be restricted to the cool, moist habitats of montane regions, although it also occurs in coastal lowlands, for example, around San Felipe in the Baja California province of northwest Mexico (personal observation). Clearly, it is not well adapted to extremely arid conditions and in this respect is similar to the scolopendromorph centipedes just considered and different from the solifugids, spiders, and scorpions on which we can now focus our attention.

The data presented in Table 6.1 indicate that the Sudanese solifugid (camel-spider), *Galeodes arabs*, has good powers of water

TABLE 6.1.
Comparative Data on Rates of Body Water Loss of Soil-Dwelling Desert Arachnids

Group	Locale	Temperature (°C)	Water Loss (% body weight hr^{-1})
Uropygi			
Mastigoproctus giganteus	United States	30	0.300[a]
Solifugida			
Galeodes arabs	North Africa	33	0.090[b]
Araneida			
Eurypelma sp.	United States	33	0.147[c]
Scorpionida			
Hadrurus arizonensis	United States	30	0.028[d]
Leiurus quinquestriatus	North Africa	33	0.030[e]
Androctonus australis	North Africa	33	0.032[e,f]
Centruroides sculpturatus	United States	30	0.091[d]
Buthus hottentotta	West Africa	25	0.105[g]
Diplocentrus spitzeri	United States	30	0.320[h]

[a]Ahearn 1970b.
[b]Cloudsley-Thompson 1961.
[c]Cloudsley-Thompson 1967.
[d]Hadley 1970b.
[e]Cloudsley-Thompson 1961a.
[f]Cloudsley-Thompson 1956.
[g]Recalculated from Toye 1970, as an average.
[h]Crawford and Wooten 1973.
Source: Compiled by the author.

retention compared with, for example, the tarantula spider, *Eurypelma*, which inhabits parts of the Sonoran desert, and the semimontane scorpion *Diplocentrus spitzeri*. The camel-spider's ability to conserve water is on a par with that of the American scorpion *Centruroides sculpturatus*, but it is less efficient in this respect than the extremely drought-resistant *Androctonus australis* and *Leiurus quinquestriatus* of North Africa and *Hadrurus arizonensis* of North America. *Galeodes* is an extremely efficient nocturnal predator, and its mobility allows it to exploit water sources that are available in its prey. It has developed a survival strategy that treads a fine line between water gain and water conservation. The same may be true of the spider *Lycosa carolinensis*, which is a soil burrower in the Sonoran desert uplands. This species cannot subsist solely on metabolic water but must tap water from the fluids of its prey, from dew and from capillary water in the soil (Moeur and Ericksen 1972).

In coping with the physical excesses of the desert environment scorpions, like all other animals, have to ensure an adequate supply of water. This they obtain for the most part from the fluids of their prey, but they will resort to drinking free water when the occasion demands. They do not have the ability, at least as adults, to take up water from unsaturated air or moist surfaces and in this regard resemble the scolopendromorph centipedes.

But it is their ability to conserve body water that has enabled scorpions to become successful desert animals. Their nocturnalism helps in this respect since this behavior ensures that their periods of surface activity are confined to the hours of darkness when atmospheric relative humidity is higher than during the day. Coupled with this feature is the possession of a relatively impermeable cuticle that restricts transpiration losses, as the data in Table 6.1 indicate. *Hadrurus arizonensis*, for example, has a marked transition temperature around 40°C to 44°C, implying that an epicuticular waterproofing lipid layer is present. These desert arthropods are large in body size, and their low surface area/volume ratio is also a factor in reducing water loss. Moreover, they have low metabolic rates, which also operate to conserve water. Their selection of sheltered microhabitats under rocks or tree bark or in soil burrows and their positioning within these refugia serve to minimize environmental stress (Hadley 1970b). They are morphologically adapted for living in these microsites by virtue of their dorso-ventrally flattened shape and their ability to dig in the soil.

The scorpions have appeared in two contexts in this chapter and the previous one: as avoiders and as tolerators. Indeed, perhaps it will be useful to make comparisons between these two main ecological groupings at this point. Avoidance is achieved in one of two ways. The first is to select montane or semimontane habitats where mesic conditions prevail. *Diplocentrus spitzeri*, for example, is common in the Peloncillo Mountains of the southwestern part of New Mexico where rainfall events are unpredictable in their timing and magnitude. However, *D. spitzeri* selects mesic, semimontane regions in this area and thereby is not exposed to the kind of moisture stress that would occur in lowland sites. This species has a relatively high transpiration rate for a desert scorpion (Table 6.1) and would not be expected to live in very arid conditions. The second avoidance mechanism is to burrow deeply in the soil where moisture conditions are more equable than at the surface. Examples of this tactic are shown by *Scorpio maurus* and *Buthus occitanus*.

Not all desert scorpions are deep burrowing or restricted to semimontane habitats, however. Some are equipped to tolerate arid conditions better than the examples just cited, and these include *Leiurus quinquestriatus*, *Androctonus australis*, and *Buthotus minax* of North Africa and the North American *Hadrurus arizonensis* and *Centruroides sculpturatus*. As is obvious from Table 6.1, the adaptions of these tolerant species are centered around the ability to conserve water by restricting transpiration rates. They also have efficient mechanisms for closing off the openings to the lung books, thereby restricting respiratory water loss.

As is well known, scorpions are venomous animals, and the venom is ejected from the tailpiece, or telson, at the end of the metasoma. This venom represents a source of water loss, and it is used, mainly for defensive purposes (Kaestner 1968). However, it may also be used to narcotize prey caught in the pedipalps. The toxicity of scorpion venom varies considerably from species to species, and there appears to be an inverse relationship between the development of the pedipalps and the strength of the venom. Species that are considered to be dangerous, because of their sting, such as *Leiurus quinquestriatus*, *Androctonus australis*, and *Centruroides sculpturatus*, have slender pedipalps (Fig. 6.2), whereas the less-feared *Pandinus imperator*, *Diplocentrus spitzeri*, and

Fig. 6.2. The North African scorpion *Leiurus quinquestriatus* in threat posture. Note the slender pedipalps. (Photo courtesy of J. L. Cloudsley-Thompson.)

140   Desert Soil Fauna

Scorpio maurus (Fig. 6.3) have heavily developed chelae on the pedipalps. There also appears to be a relationship between the toxicity of scorpion venom and habitat. The species that live in the most arid environments produce the most toxic venom. The toxicity of the venom is a function of its concentration, and this implies a tactic for water conservation. Scorpions living in arid regions produce concentrated poisons that immobilize their prey quickly, effectively, and with a minimum of water loss.

Amphibians

These vertebrates can be considered primarily as avoiders in desert systems by virtue of their burrowing behavior, nocturnalism, and seasonality. However, desert amphibians exhibit a suite of adaptations that can more properly be considered as tolerance mechanisms. These adaptations are particularly well developed in those species of the spadefoot toad genus *Scaphiopus* that live in the more arid sites; it is on these desert anurans that much attention has been focused.

Fig. 6.3. The scorpion *Scorpio maurus* in the Israel desert. Note the heavily developed pedipalps. (Photo courtesy of J. L. Cloudsley-Thompson.)

In the first instance *Scaphiopus* practices ureotelism and can tolerate high levels of urea in the body fluids (McClanahan 1975). The rate of urea production in these anurans is a function of soil water potential; when environmental moisture is scarce, urea production is high (McClanahan 1972). In this way the water potential between the toad and the soil is decreased, water loss across the body surface is minimized, and uptake of water from the soil is favored (McClanahan 1967; Ruibal, Tevis, and Roig 1969; Walker and Whitford 1970). Water uptake from the soil is further facilitated by orientation behavior within the burrow, and Hillyard (1975) found that *S. couchii* could absorb soil moisture three times more rapidly by changing its position in the burrow. The effect of these positional changes was to apply the skin surface to new sources of soil moisture once old ones had been exhausted.

Scaphiopus couchii is only surface active during summer rains. At this time there is plenty of free water that can supplement the preformed water in the body fluids of the prey, and toads rehydrate readily under these conditions. McClanahan (1967) reports that *S. couchii* does not rehydrate more rapidly than other anurans and, indeed, compares unfavorably in this regard with the North American *Bufo punctatus* and the Australian *Neobatrachus*; but this may be one aspect of the desiccation tolerance of *S. couchii*. On the other hand, at the extreme western edge of its range in the southwestern United States *Scaphiopus couchii* faces a problem. Here in southeastern California summer rains are uncertain (this is mainly a winter rainfall desert; see Chapter 5), and toads may have to remain dormant in the soil for perhaps two years. During this period there is very little opportunity for water gain from the environment; uptake of water across the skin is a possibility that has already been discussed. The other possibility is the utilization of metabolic water. McClanahan (1967) examined coelomic fat deposits in emerging and foraging toads, coupled this with laboratory-based determinations of respiratory quotients, and concluded that the toads lay down fat reserves during the feeding season and catabolize these during the subsequent inactive subterranean phase. The mobilization of lipid during periods of water stress provides a source of energy and metabolic water. Whitford and Meltzer (1976) estimated that juvenile *Scaphiopus hammondii* utilized 40 percent of their total fat reserves in a six-month period of dormancy and at this rate of depletion could survive for an additional six-month period. Seymour (1973) estimated that lipid metabolism in dormant *S. couchii* and *S. hammondii* provided about 50 percent of the energy requirements during periods of water stress. This author also examined the fat bodies of emerging toads and noted that in some cases these were large enough to enable the toads to survive periods of two or more years of starvation.

Another source of metabolic water is the catabolism of protein. As we will see later, this is a major source of water for the estivating desert snail *Bulimulus dealbatus*, and it may also make a contribution to the water relations of desert toads. This is suggested by the accent placed on ureotelism and the ability of *Scaphiopus* spp. to tolerate high levels of urea in their body fluids—a by-product of this metabolism that is put to good effect in reducing water potentials between the animal and its environment (see earlier).

Despite these various avenues of water gain, desert amphibians are subject to water stress in arid environments. They needed to develop some mechanisms for water conservation, and this they have done. First, the reduction in the water potential between the body wall and the soil will curtail water loss, as already indicated. Second, protection from water loss is provided by the retention of cast skin as a cocoon in the case of the South American *Ceratophrys* and the Australian desert frogs (see Chapter 3). Third, the presence of wax esters in the skin is a common feature of desert anurans (Shoemaker 1980), and this provides another method of waterproofing. Fourth, the ability to detoxify ammonia and become ureotelic has important consequences as far as the conservation of body water is concerned. Urea is much less toxic than ammonia, and the ability to tolerate high levels of this substance in the body fluids obviates the need for its elimination in solution from the body. *Scaphiopus*, for example, can store up to 300 mM/l of urea in its body fluids during dormancy (McClanahan 1967; Shoemaker, McClanahan, and Ruibal 1969; McClanahan 1975). As a consequence of this storage and the antidiuretic response of kidney and bladder (Hillyard 1975), a shutdown of kidney functioning becomes possible (Shoemaker 1980), and this seems to be a fairly widespread phenomenon among desert anurans. Fifth, dormancy in *Scaphiopus* and *Ceratophrys* is characterized by a reduction in the metabolic rate of about 25 percent, and this will result in a curtailment of respiratory water loss. Finally, desert anurans have a large urinary bladder, which serves as an important water reserve. The fluid in this bladder is relatively dilute (McClanahan 1967), possibly as a result of high urea production and/or reduced glomerular filtration rates. The adaptive value of this dilute bladder fluid will shortly become apparent.

Although they have mechanisms for water gain and water conservation, anurans are very much at risk in the hot desert environment. Mortality among overwintering juveniles is particularly high (Creusere and Whitford 1976; Whitford and Meltzer 1976), and this is probably due to dehydration and/or depletion of fat reserves. On the other hand, adult *Scaphiopus* can dehydrate to the extent of losing 40 to 60 percent of their body weight and recover (Thorson and Svihla 1943; McClanahan 1967; Whitford and Meltzer 1976). Dehydration, in the normal course of events, involves a loss of water from the blood and a consequent decrease in plasma volume. *Scaphiopus couchii* appears to be able to maintain plasma volume levels during dehydration better than such semiaquatic species as *Bufo marinus*, although this capacity is no more greatly developed than in other terrestrial anurans. The key to the mechanism that provides for constant plasma volume evidently resides in the dilute bladder fluid. Movement of this fluid from the bladder to body fluid compartments ensures that plasma water levels are "topped up," and this will be maintained as long as the bladder water supply lasts; after this point has been reached, water loss occurs from interstitial fluids, not from the blood (McClanahan 1967).

Lizards

In contrast to the amphibians we have just been considering, desert reptiles, in the main, have a much more tenuous connection with the soil, and it is beyond the brief of this book to attempt an exhaustive survey of their water relations. Some points deserve a passing mention, however.

First, the majority of lizards are carnivores and can utilize preformed water in their prey. Metabolic water, produced by the oxidation of fats, may also be of considerable survival value; the body tissues of lizards are rich in fat deposits that can be mobilized in times of water deprivation. Alternatively, there is the possibility that small-sized species of desert reptiles may exploit the moisture sources provided by dew formation (Mayhew 1968).

Second, water loss through transpiration is reduced by the possession of a dry, scaly skin. Rates of cutaneous water loss in the lizard *Uta stansburiana hesperis* are about 100 times less than in desert anurans (Claussen 1969); and Chew and Dammann (1961) reported that desert reptiles lose water at a rate that is 10 times lower than that of desert rodents. Clearly, desert reptiles compare very favorably with the other vertebrates we are considering with regard to cutaneous water loss, particularly when it is realized that 39 percent of the total body water loss occurs, at 30°C, along this pathway in *Uta stansburiana* (Mayhew 1968).

Third, respiratory water loss is curtailed by behavioral adaptations designed to avoid exposure to high temperatures. Fossorial lizards such as *Uromastix aegypticus, U. loricatus, U. hardwickii, Palmatogecko* sp. go underground, as does the skink *Egernia kintorei*, which constructs "family" burrows (Mayhew 1968). Some lizards, such as *Dipsosaurus dorsalis*, make use of the arboreal escape route, climbing into the foliage of the creosote bush to avoid high ground surface temperatures. These behavioral traits are, essentially, concerned with body temperature regulation and as such are considered in more detail in the following chapter. However, it is relevant to mention them at this point for two reasons: (1) lizards are exposed to temperature stress on occasions (they are, after all, predominantly day-active), and at these times they will pant; in doing so they lose water through the respiratory pathway; and (2) water loss through the cutaneous pathway will increase owing to the activation of evaporative cooling mechanisms. Behavioral patterns that remove reptiles from situations of temperature stress will obviously reduce water loss along these pathways.

Fourth, water loss via the excretory system is reduced to a minimum by uricotelism. The formation of insoluble urates as the end products of nitrogen metabolism means that these waste materials can be eliminated from the body with a minimum of water loss. In *Uromastix hardwickii*, for example, water resorption from the urine occurs to such an extent that waste products are eliminated as a hard pellet (Mayhew 1968).

Marsupials

The insectivorous marsupials of the Australian deserts are primarily avoiders and alternate periods of nocturnal activity with daytime torpor, at least in the cases of *Dasycercus cristicauda* and *Sminthopsis froggatti*. As such, they will experience little or no water stress. However, the fat-tailed marsupial mouse *S. crassicaudata* remains active during the diel cycle and may have to call upon the water obtained from fat metabolism during prolonged periods of activity.

SURFACE DETRITIVORES AND HERBIVORES

In contrast to the surface-active predators that we have just been considering, soil-based detritivores and herbivores that forage on the surface are not generally nocturnal in habit. To be sure, woodlice, certain insects, and many granivorous rodents are active at night, but daytime activity is shown by desert millipedes, many ants, tenebrionid beetles, and snails. This activity is often confined to the rainy season, and life cycles are geared to this, as we have already seen. This is an avoidance mechanism, but the dormant state that is assumed during unfavorable periods of the year has attendant physiological problems that must be overcome. This is well illustrated by the North American desert millipede *Orthoporus ornatus*, which has been studied in detail by Crawford (1972, 1978).

Millipedes

The alternation of a period of dormancy with one of feeding activity, subterranean and surface, respectively, presents *Orthoporus ornatus* with two different types of problem each year, as far as its water relations are concerned. During the feeding period, water is obtained from the vegetation on which the millipede feeds and also by oral and anal drinking (Crawford 1972). It may also be taken up actively across the body surface from saturated air or produced metabolically (Crawford 1978). During subterranean dormancy, the avenues of water gain are much more restricted since preformed water in the food is not available, and the dormant animal does not actively imbibe free water. But there are other potential sources that may be exploited at this time, such as metabolic water, and reserves contained in the gut and hemolymph. Crawford (1978) considered each of these possibilities in turn. He calculated that the production of metabolic water was so low as to be insignificant in the water budget of *O. ornatus*. The reserves of water stored in the gut and hemolymph could satisfy about 50 percent of the water requirements of the cuticle-tissue compartments, particularly during the early part of the dormant period. However, the gut and hemolymph act as temporary reservoirs, and they become exhausted midway through the dormancy period, by which time *O. ornatus* is in a state of minimal hydration. Beyond this point (which occurs in November in the Chihuahuan and Sonoran deserts) and until the onset of rains (in the following July), cuticle-tissue water increases, probably due to active transcuticular uptake from the soil. The mechanism underlying this uptake has not been elucidated, but it must involve a reduction in the water potential across the body surface; here, possibly, there is an analogue of the condition that occurs in *Scaphiopus* (see earlier).

These studies by Crawford, complemented by those of Riddle, Crawford, and Zeitone (1976) and Pugach and Crawford (1978), show that *O. ornatus* regulates its water balance during both its dormant and feeding periods. It does this not only by exploiting environmental moisture from a variety of sources but also by conserving its body water in various ways. In the first instance it has the lowest transpiration rate of any millipede studied to date (Table 6.2). The mechanism whereby *O. ornatus* reduces transpiration losses has not been fully explained. Experiments by Crawford (1972) indicate that large specimens exhibit a significant increase in cuticular permeability at temperatures between 35°C and 40°C; this does not occur, apparently, in small specimens (Fig. 6.4). There are interesting similarities and differences here with the whip scorpion *Mastigoproctus giganteus* studied by Ahearn (1970b). Transpiration rates are very similar at 30°C (0.30 percent for *M. giganteus*; 0.24 percent for *O. ornatus*). *M. giganteus* has a transition temperature that lies between 35°C and 40°C, which Ahearn interpreted as an indication of the presence of a waterproofing barrier in the cuticle. However, this barrier, if it is present, occurred in small rather than large specimens (see earlier in this chapter). Crawford (1972) found that transpiration rates were inversely correlated with body weight and directly related to surface area between 35°C and 40°C in large specimens of *O. ornatus*. These rates also increased with the death of these animals. Such observations strongly suggest a loss of transpiratory control across cuticular or respiratory surfaces at high temperatures or on death. That this increased water loss may be due, at least in part, to enhanced

TABLE 6.2.
A Comparison of Body Water Loss from Various Species of Millipedes in Dry Air at 30° C

Species	Habitat	Water Loss (mg cm^{-2} hr^{-1})
Tachypodoiulus niger	Mesic	4.3[a]
Oxydesmus platycerus	Mesic	3.8[a]
Ophistreptus sp.	Mesic	7.2[a]
Orthoporus ornatus	Xeric	0.25 (max)[b]

[a]Cloudsley-Thompson 1959.
[b]Crawford 1972.
Source: Data from Crawford 1972.

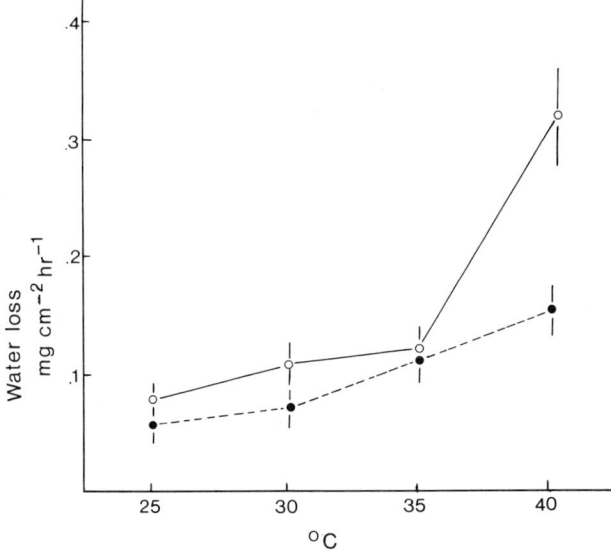

Fig. 6.4. Comparative rates of water loss from large and small specimens of *Orthoporus ornatus* in dry, moving air for 12 hours: open circles: large specimens; closed circles: small specimens. (From Crawford 1972.)

locomotor activity between 35°C and 40°C is suggested by aktograph studies (Fig. 6.5). Just how much of the water transpired at these high temperatures is lost through the cuticle, as opposed to the spiracles, is unclear. Certainly the cuticle of large specimens has a transition temperature that would indicate the breakdown of some kind of lipid water barrier. On the other hand, Crawford (1972) found that rates of water loss were significantly lower in coiled specimens (in which the spiracles and legs are minimally exposed) at 40°C than in specimens that were held straight. This difference could not be detected at 30°C, which suggests that the cuticle is the main avenue for water loss at the lower end of the 30°C to 40°C range but that a failure of the spiracular closing mechanism may occur at more stressful temperatures.

Coiling is a behavioral device that many temperate millipedes exhibit, and it is generally regarded as a means of avoiding desiccation. However, *O. ornatus* assumes the coiled position, partly or completely, at temperatures below 35°C irrespective of the ambient humidity (Crawford 1972). It also coils during the dormant period in the soil, and this may be more a defensive posture than anything else.

During dormancy transpiration losses will be minimal since

148 Desert Soil Fauna

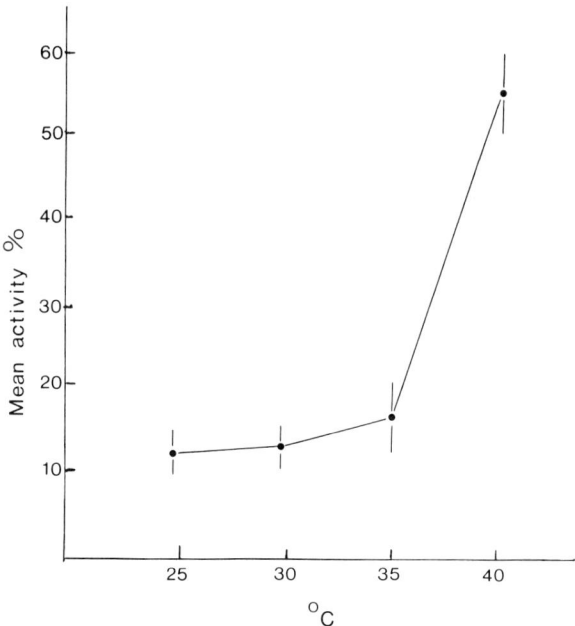

Fig. 6.5. Relative activity of medium-sized *Orthoporus ornatus* in dry, moving air for 30 minutes. (From Crawford 1972.)

the hibernacula in which the animals live are protected from climatic extremes. However, there is one exception as far as *O. ornatus* is concerned, and that is during the May/June period when the populations of this millipede studied by Crawford (1978) underwent their annual molt. At this time there is a decrease in total water content, possibly associated with the new, soft cuticle and/or the production of molting fluid. However, Crawford's data also indicate that the relative amounts of water in the gut and hemolymph increase over the molting period. This may be a way of limiting body water loss by shifting this commodity from cuticular sites deeper into the body core.

Other avenues of water loss exist, at least potentially, when these desert millipedes are active on the soil surface; these are fecal water loss and loss due to secretion from the repugnatorial glands. Crawford (1978) estimated that fecal water loss comprised less than half of the total water loss during the feeding period. Losses from the repugnatorial glands may be considerable (although no estimates are available) since the secretions from these glands not only serve a defensive purpose but also serve to stimulate swarming behavior when the millipedes are not feeding. However, if desert millipedes adopt the same type of strategy as desert scorpions for restricting water loss along this pathway, it could be expected that their repugnatorial secretions would be more concentrated than those of their relatives in moist temperate regions.

Clearly, *Orthoporus ornatus* is unusually well adapted (for a millipede) to life in an arid environment, and its exploitation of desert habitats reflects this fact. Other millipedes, such as the polydesmoid *Habrodesmus falx* and *Oxydesmus* sp. of West Africa, are very much less resistant to desiccation (Toye 1966a) and as a consequence are confined to semiarid regions around the periphery of deserts. There are parallels here among the woodlice that show varying degrees of adaptation to aridity. It is to this group that we can now turn our attention.

Woodlice

These isopods occur in deserts generally where there are moist microsites available for colonization: in the shaded margins of arroyos, on the margins of playa lakes, in the shelters provided by rocky debris on scree slopes and gravel plains, and under accumulations of organic debris. This selection of moist microhabitats together with their nocturnal and/or burrowing habit could identify these arthropods as avoiders, rather than tolerators, of arid environments. However, this is only a small part of the story, for woodlice in general are preadapted behaviorally, anatomically, and physiologically to cope with arid conditions. The success of the desert-dwelling species lies not so much in their ability to evolve new and more specialized adaptions to life in arid environments but rather to perfect the physiological, structural, and behavioral traits inherited from their, presumably, mesic ancestors.

In the first instance nocturnal foraging behavior can maximize water gain from the environment. Woodlice are mainly detritus feeders, and the particles of organic debris that form their diet will have a maximum water content at night, owing to condensation. *Hemilepistus reaumuri*, for example, is able to tap moisture not only from its moist food but also by drinking from dew condensates (Shachak 1975). This species has a crepuscular activity pattern and may take advantage of dew formation during the early morning hours. Woodlice will also undoubtedly imbibe free water, but there is no evidence to suggest that transcuticular absorption of moisture occurs directly from the atmosphere in xeric species (Warburg 1968). Instead, the emphasis is placed on conserving body water, and this is done in various ways, all of which are designed to reduce transpiration losses across the cuticle and through the respiratory system.

Despite some early statements to the contrary, there is evidence for the presence of a water barrier in the isopod cuticle, but this is located in the endocuticle rather than in the epicuticle (Bursell 1955). All desert woodlice are much more efficient than their mesic relatives in limiting the amount of water loss across the body surface in dry air (Edney 1968b; Warburg 1968). This may be due to a more efficient endocuticular water barrier, but it may also be a reflection of a much more efficient control of respiratory water loss.

The principal respiratory structures of isopods are five pairs of flattened, platelike pleopods located ventrally on the abdomen. Species that inhabit moist environments, such as *Oniscus asellus*, use the entire surface of these pleopods for respiratory exchange. More xeric species restrict respiratory exchange to invaginated portions of these pleopods, known as pseudotracheae. The development of pseudotracheae appears to be related to the degree of aridity to which a species may be exposed. Desert-dwelling isopods have pseudotracheae on all five pairs of pleopods, whereas the less tolerant *Armadillidium vulgare* of semiarid and mesic environments has pseudotracheae on only two pairs of pleopods.

These twin adaptations of restricted cuticular and respiratory water loss find their culmination in the desert woodlouse of the southwestern United States, *Venezillo arizonicus*. This species has the lowest transpiration rate of all desert woodlice and appears to have an epicuticular water barrier since there is a marked transition temperature between 35°C and 40°C (Warburg 1965c). This barrier, if it exists, is an innovation—a new and specialized feature evolved in *V. arizonicus* as an adaptation to extreme physical conditions. Transcuticular water loss is further restricted by the absence of tegumental glands in this species (Warburg 1968). The function of these glands in mesic forms has not been clearly established; they may produce repugnatorial secretions or they may serve to keep the pleopods moist. In any event their absence in *V. arizonicus* and, indeed, in other desert isopods such as *Armadillo officinalis* and *A. albomarginatus*, can be interpreted as a means for reducing water loss across the general body surface. Here there are parallels with desert rodents that lack sweat glands (see below). The absence of tegumental glands is a feature of those woodlice species that can conglobate. *V. arizonicus, A. officinalis*, and *A. albomarginatus* can conglobate completely, even withdrawing the antennae. This habit of rolling the body into a ball reduces respiratory water loss across the pleopods since these structures are effectively sealed off, within the enfolded body, and removed from the potentially desiccating effects of the external environment.

Porcellio olivieri can neither conglobate nor burrow and evidently is less well equipped than the species just mentioned for life in arid environments. However, it can switch from negative to positive photoreaction at high temperatures, and this may enable it to locate cooler, more moist microenvironments. This behavior is more properly regarded as an avoidance tactic, but *P. olivieri* may also combine this with evaporative cooling, albeit to a limited extent. This may be sufficient to postpone heat torpor long enough for shelter to be found. The species is common in the Mediterranean deserts where it lives in wadis periodically inundated by rains. Here moist microsites under rocks would not be too infrequent. Indeed, Kheirallah and Awadallah (in press) have reported that peak densities of this isopod are closely correlated with seasonal rainfall.

Evaporative cooling is a luxury few desert arthropods can afford because it depletes precious body water stores. The Armadillidae, which include the best adapted desert isopods, do not indulge in this method of thermoregulation (Warburg 1968). Their recipe for success in hot deserts is a combination of a low transpiration rate, the ability to conglobate, and a nocturnal pattern of activity. Warburg (1968) considers *Hemilepistus reaumuri* as a porcellionid that does not use evaporative cooling, but this is in contradiction to Edney's (1958, 1968b) findings. Be this as it may, it seems clear that *H. reaumuri* relies mainly on its low transpiration rate, its ability to burrow, and its sociality to survive in hot deserts.

To summarize briefly, there are at least three different strategies for combating aridity among the xeric isopods we have been considering. Distinctions can be made, in this respect, among the Armadillidae on the one hand, *Porcellio olivieri* on another, and finally, *Hemilepistus reaumuri*. Each of these groups has selected a different combination of features, and the pattern that emerges is depicted in Table 6.3.

Although we are principally concerned with the adaptions of animals to extremely arid environments, it is instructive to shift attention from time to time to the margins of these environments—to the edges of the desert where the transition zone between mesic and xeric has its own peculiar fauna. There are good examples of this among the isopods.

Armadillidium vulgare is a widely distributed isopod in grassland and semiarid regions. Although it is often designated as an inhabitant of mesic environments, it has a relatively low transpiration rate and can conglobate, although incompletely since it cannot

TABLE 6.3.
Survival Strategies Employed by Different Species of Desert Woodlice

	Porcellionidae		Armadillidae		
Tactic	Porcellio olivieri	Hemilepistus reaumuri	Armadillo officinalis	Armadillo albomarginatus	Venizillo arizonicus
Restricted water loss	+	+	+	+	+
Conglobating	−	−	+	+	+
Burrowing and sociality	−	+	−	−	−
Evaporative cooling	+	?	−	−	−
Photoreaction	(−)*	(+)*	−	−	−

*Reversed at high temperatures.
Source: Table based on Warburg 1968.

withdraw its antennae. Furthermore, it can retreat into deep soil crevices during periods of drought (Paris 1963). Studies on populations of this species in Arizona have shown that transpiration rates are much lower than those of populations of the same species in England. Cloudsley-Thompson (1969) suggested that this may be the result of different selection pressures over a long period of time, or an acclimation effect during the lifetime of an individual, or a combination of both of these factors. He also showed that populations of the semixeric *Metaponorthus pruinosus* in North Africa have a similar kind of pattern, with specimens from the Suden possessing considerably lower transpiration rates than Algerian specimens. *M. pruinosus* is common in gardens and orchards around Khartoum where it occurs with another isopod, *Periscyphis jannonei*. Both of these species have low transpiration rates, that of *P. jannonei* being almost as low as that of *Hemilepistus reaumuri*.

It will be obvious that these isopods that live around the margins of hot deserts have at least some of the attributes necessary for survival in arid environments. That they are less well adapted, in this respect, than the truly xeric *Hemilepistus reaumuri* and *Venizillo arizonicus* emphasizes the extremely narrow dividing line between survival and extinction in hot deserts. Thus, even though *Armadillidium vulgare* can conglobate, it has a much higher transpiration rate than *Hemilepistus reaumuri*, which cannot (Table 6.4). The latter has more pseudotracheae than *A. vulgare*, however, and can burrow more efficiently. *A vulgare* moves deeper in the soil, mainly by making use of existing fissures and rodent burrows (Paris 1963). It also congregates, particularly in the juvenile stage, in moist areas around the bases of grasses and shrubs, but these aggregations do not have the same kind of survival significance as they do in *H. reaumuri*. Like the other semixeric isopods we have been considering, *A. vulgare* relies to some extent on regaining the water lost through transpiration from its immediate environment. In contrast *H. reaumuri* and the Armadillidae rely on conserving water—the hallmark of true desert animals.

Snails

Desert snails such as the Middle Eastern *Sphincterchila zonata* and *Trochoidea seetzeni* and the North American *Rabdotus schiedeanus* take in water with their food, and they feed only during the rainy season (Schmidt-Nielsen, Taylor, and Shkolnik 1972; Riddle 1975). At other times of the year they assume a dormant state that is characterized by a relatively high lethal temperature, a very low metabolic rate, and a very low rate of water loss from the body. Riddle (1975), for example, showed that the metabolic rate decreased with decreasing relative humidities in *R. schiedeanus*, and this was associated with reduced activity during unfavorable periods. This is probably a good way of conserving energy reserves—apparently, it is also utilized by the scorpion *Hadrurus arizonensis* (see earlier in this chapter and also Hadley 1970b). A reduction in the metabolic rate is also a feature of the early stages of estivation in *Bulimulus dealbatus*. This North

TABLE 6.4.
Rates of Transpiration of Some Isopods

Species	Transpiration Rate (mg cm^{-2} hr^{-1} mm Hg^{-1} × 10^3 at 20°C to 30°C)	Habitat
Porcellio scaber	110	Hygric
Armadillidium vulgare	85	Mesic
Hemilepistus reaumuri	23	Xeric
Venezillo arizonicus	15	Xeric

Source: Data taken from Edney 1968b.

American species, like *Otala lactea*, which will be considered in more detail shortly, is semixeric and, while it may not be completely desert-adapted, has some of the features of arid zone mollusks. Horne (1973), for example, has reported that respiration rates in *B. dealbatus* declined to 16 percent of resting active levels three days after the onset of estivation. During estivation *B. dealbatus* catabolized body tissues (protein: 57 percent; carbohydrate: 35 percent) to produce much needed water and energy, and this is also a feature of the estivation metabolism of *Sphincterochila zonata* (Schmidt-Nielsen, Taylor, and Shkolnik 1971). The metabolism of protein produces urea as a by-product; as we have already seen in the case of a desert spadefoot toad *Scaphiopus*, high urea concentrations in body fluids can reduce water potentials and, thereby, water loss across the body surface. However, Horne (1973) suggests that the secretion of epiphragms may be more important in reducing water loss in *B. dealbatus* than any mechanisms that are physiologically designed to reduce water loss. Clearly the avenues of water gain are limited for these desert snails, and the conservation of body water during the dormant period assumes a special significance, particularly since this dormancy, at least in the case of *S. zonata*, is often spent on or near the soil surface.

S. zonata is a relatively small snail, and its size could be considered a disadvantage because of the high surface area/volume that this entails. However, its shell is significantly thicker than that of mesic helicids such as *Helix aspersa* and *Otala lactea* (Machin 1967). It also has a smaller shell aperture, relative to its size, than *H. aspersa* and *O. lactea* and a much thicker epiphragm. All of these features combine to reduce body water loss to a level that would allow *S. zonata* to survive for several years in a dormant condition (Schmidt-Nielsen, Taylor, and Shkolnik 1972). Indeed, Shachak, Chapman, and Orr (1976) calculated that *S. zonata* has a water uptake of 406 μl day^{-1} and an energy uptake of 227 cal day^{-1} during its active period (20 to 25 days). In contrast the water loss of an estivating snail is only of the order of 2.3 μl day^{-1}; energy losses during dormancy are similarly of a very low order of magnitude (1.9 cal day^{-1}). The sympatric *Trochoidea seetzeni* shows similar powers of water retention.

However, there is probably much more to this story than just the thickness of the shell, the size of its aperture, and the thickness of the epiphragm. Detailed studies on water regulation in estivating *Otala lactea* indicate as much (Newell and Machin 1976; Appleton, Newell, and Machin 1979; Newell and Appleton 1979). *O. lactea* is more properly considered as a semixeric species, but there is no reason to believe that its physiology differs in any significant respect from that of xeric species.

When it estivates, *O. lactea* withdraws into its shell and plugs the aperture with mantle collar epithelium. Water loss from this epithelium is comparable with that across insect cuticle in estivating snails. This remarkable ability to control water loss from soft, moist epithelial tissue is evidently militated by the development of a vesicular water barrier in the apical portions of these epithelial cells. Here osmotic gradients of chloride and potassium ions are strongly developed in estivating snails (compared with active ones), and these gradients prevent water from being mobilized from underlying tissue. Perhaps there is more to this story still, for research is continuing (Peter Newell 1981: personal communication), and surely it would be rewarding to apply this approach to truly desertic species such as *Sphincterochila zonata*, *S. prophetarum*, and *Rabdotus schiedeanus*.

Insects

In this section on insects we can concentrate on four groups with drought-tolerant representatives: termites, roaches, tenebrionid beetles, and ants. Passing mention can also be given to the hexapod Thysanura, although these arthropods have been less intensively studied than the other groups just mentioned. Virtually nothing is known of the drought tolerance of other groups in hot deserts: psocopterans, japygids, thysanopterans, and dipteran and coleopteran larvae. These must, unfortunately, be excluded from further discussion, but perhaps it is not a vain hope that they will be included in future research.

Desert termites are only of marginal relevance to the theme developed here. They have a permeable cuticle that renders them very susceptible to desiccation (Moore 1969). Some species can survive for only a few minutes at temperatures in the range of 34°C to 35°C and relative humidities of 0 to 4 percent (Collins 1969). However, the dry-wood termites of the family Kalotermitidae are more tolerant and can survive for days or weeks under such conditions. Dry-wood termites are not uncommon in hot deserts, and their ability to tolerate arid conditions is just one aspect of their survival strategy. Of particular relevance is the fact that this family of termites is the only one capable of resorbing water from the feces before they are voided (Bouillon 1970). This is yet another illustration of the fact that successful exploitation of arid environments can only occur through efficient water conservation.

The desert cockroach of the New World, *Arenivaga*, has developed mechanisms for maximizing water gain from the environment and minimizing water loss. The wingless nymphs and adult females (see Fig. 2.13) have the unusual ability to absorb water vapor from the atmosphere at relative humidities above 82.5 percent at temperatures between 10°C and 30°C (Edney 1966a); winged adult males cannot do this. This uptake involves the movement of water from a humidity of 82.5 percent to the equilibrium humidity of *Arenivaga* hemolymph (99 percent)—in other words, against an osmotic gradient of more than 100 atmospheres. This special physiological mechanism for regaining water from the environment is unusual because although it is employed by hard ticks, no other free-living desert invertebrate, apart from the thysanuran *Ctenolepisma terebrans* of the Namib (Edney 1971), has been recorded as having this capability. The precise mechanism employed by *Arenivaga* is quite unique, as we can now see.

The site of absorption of atmospheric water vapor in such insects as the flour beetle (*Tenebrio molitor*) and the firebrat (*Thermobia domestica*) is the rectum and its associated structures (Noble-Nesbitt 1970a, 1970b; Dunbar and Winston 1975; Machin 1975, 1979). However, these are not desert insects, and there is no reason to assume that *Arenivaga* would follow their example; indeed, it does not. O'Donnell (1977, 1978) has shown, in an elegant manner, that the site of water vapor absorption in *A. investigata* is located at the anterior, rather than the posterior, end of the body. He used wax-blocking experiments to rule out the rectum as a source of water uptake. He then identified the mouth region and in particular a pair of sacs (ventrolateral diverticula of the hypopharynx) that could be extruded from the mouth of nymphs and adult females to absorb environmental moisture. These hypopharyngeal sacs, or bladders, are densely packed with fibrils to increase the surface area and are moistened by a fluid secretion, probably originating from the salivary glands, which may be hygroscopic and which may provide sites for the condensation of water vapor. This condensed water is conveyed via the hypopharynx and esophagus to the crop and thence to the hemolymph.

Arenivaga investigata also has some specialized features associated with water conservation. Its habit of burrowing in sand exposes the cuticle to abrasion that increases permeability. However, it has the capacity for repairing an abraded cuticle in the very short time of one or two days (Edney 1966a). *Arenivaga* has also produced an unusual solution to a problem common to all desert invertebrates that experience periods of dehydration alternating with periods of rehydration—namely, osmoregulation. The dehydration process involves the loss of water from the hemolymph and tissue fluids that, as a consequence, become more concentrated. On rehydration this process goes into reverse and these fluids become more dilute. The desert woodlice that we considered earlier in this chapter apparently do not have the ability to osmoregulate but, rather, tolerate a wide range of hemolymph concentrations (Edney 1968a). *Arenivaga* does osmoregulate by withdrawing osmotically active substances from the hemolymph as desiccation proceeds, although the mechanism whereby this is done is as yet unknown.

The third group of surface-active invertebrates that we need to consider in this section are the beetles belonging to the family Tenebrionidae (Fig. 6.6). Desert tenebrionids are mainly herbivor-

Fig. 6.6. Threat posture by a tenebrionid from the Mojave desert. Note the completely fused elytra. (Photo by G. Ott.)

ous or saprophagous as larvae, whereas the adults adopt a scavenging habit, feeding on seeds and dead organic material of plant or animal origin (see Chapter 2). Personal observations on Mojave desert tenebrionids indicate that they can subsist for several weeks without access to free water. This suggests that their moisture requirements may be satisfied by the moisture they take in with their food, although this does not preclude the possibility that they will drink free water when this is available. However, they do not appear to have the capability to absorb water vapor from unsaturated air in the way that *Arenivaga* can (Ahearn 1970a). This is probably due to an impermeable cuticle that also acts to reduce water loss from the body. Indeed, the success of the Tenebrionidae in deserts can be attributed mainly to their ability to conserve body water. To understand how they do this, we must recapitulate the potential avenues of water loss.

Total body water loss in desert tenebrionids is divisible into three compartments: fecal water loss, total transpiration loss, and quinone release. Fecal water loss incorporates not only water passing out of the alimentary canal but also excretory water. Insects are noted for their ability to resorb water from the feces via rectal glands and from excretory products across the wall of the Malpighian tubules, so it seems unlikely that the fecal pathway represents a major drain on total body water. Indeed, Ahearn (1970a) could detect no positive contribution to total body water loss from fecal production in the tenebrionid *Eleodes armata* under experimental conditions. Total transpiration losses are divided between cuticular transpiration and respiratory water loss. As Hadley (1979) points out, cuticular transpiration predominates at low temperatures, while increased ventilation at higher temperatures increases respiratory losses. The combined transpiration from both these avenues can account for more than half the total body water loss in *Eleodes armata* (Ahearn 1970a). The remainder is lost through the production of repugnatorial quinones, although it must be emphasized that water loss through this avenue is likely to be higher under laboratory conditions, where animals experience stress, than in the wild. Already we have seen that the production of repugnatorial quinones in millipedes and venoms in scorpions represents a potential drain on body water, but this is notoriously difficult to quantify. There are indications that arid-adapted scorpions produce more concentrated venom than their counterparts in cooler, more moist regions, but it is not known if this applies with equal force to the quinone secretions of desert tenebrionids. A summary of Ahearn's (1970a) findings is given in Fig. 6.7.

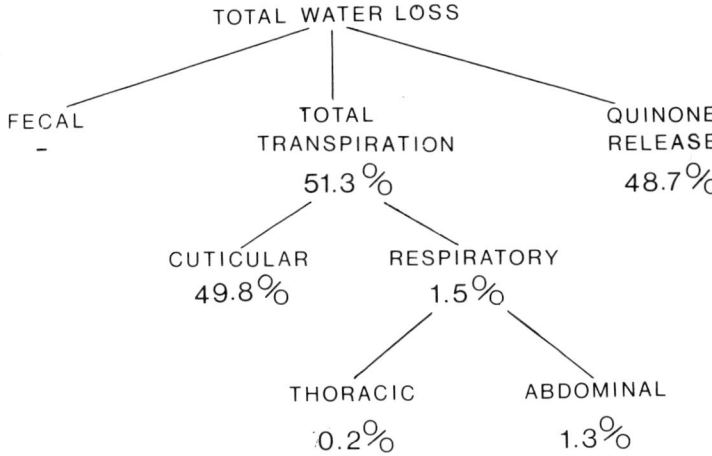

Fig. 6.7. Components of total body water loss in the desert tenebrionid *Eleodes armata* at 25°C and near 0 percent relative humidity. (From Hadley 1979, after Ahearn 1970a.)

The data given in this figure clearly indicate that conservation of body water must be effected by the judical use of the quinone defense mechanism and by minimization of cuticular transpiration. We can only speculate about the former, but there are some concrete facts relating to the latter. In the first instance desert tenebrionids are relatively large insects, and their low surface area/volume ratio will act to limit water loss across the general body surface. Second, these insects are flightless, and the two wing covers (elytra) are fused to form a dorsal shield that completely covers the abdomen (see Fig. 6.6). This shield is strongly convex and encloses a subelytral space that opens to the exterior by a restricted aperture above the terminal part of the abdomen (Fig. 6.8). This space represents a water-conserving device, since the abdominal spiracles open into it, and respiratory moisture collects here. The high relative humidity of the subelytral air reduces the vapor pressure gradient across the abdominal wall and the spiracles, and this provides an effective buffer against the desiccating effects of the external atmosphere. Just how effective this is has been demonstrated by Cloudsley-Thompson (1964), working on North African tenebrionids, and Ahearn and Hadley (1969), studying North American species. These authors showed that water loss from experimental animals increased by an order of magnitude when the elytra were punctured or removed completely. Third, the tenebrionid cuticle shows a marked transition temperature, indicat-

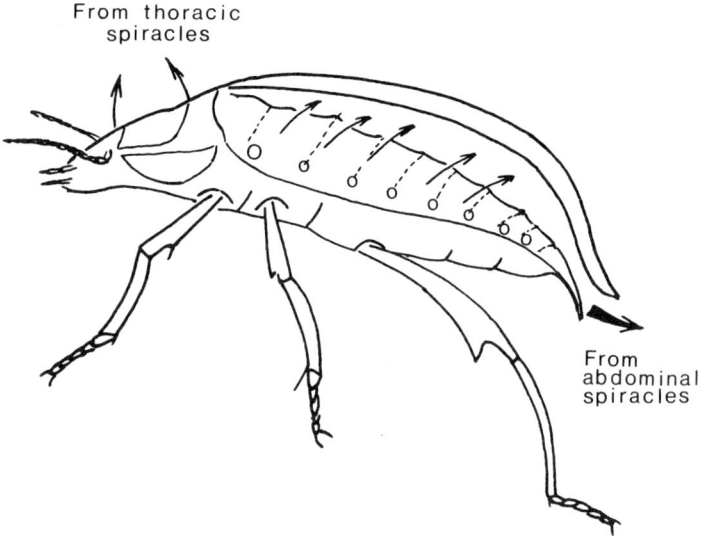

Fig. 6.8. Avenues of respiratory water loss in *Eleodes armata*. (From Ahearn 1970a.)

ive of the presence of an epicuticular lipid water barrier. This transition temperature varies from species to species and is directly related to the mode of life. This point is well illustrated by Ahearn's (1970a) study of three North American tenebrionids: *Eleodes armata*, *Centrioptera muricata*, and *Cryptoglossa verrucosa*. The first species is active mainly during the September-October period in the Sonoran desert when environmental conditions are the most favorable. It has a transition temperature around 40°C. *C. muricata* and *C. verrucosa* are summer-active beetles and as such experience conditions that are hotter and drier than those to which the active *E. armata* is exposed. That they are better adapted to cope with these conditions is reflected in their transition temperatures: 47°C in the case of *C. muricata* and 50°C for *C. verrucosa*. This gradient of transition temperatures is paralleled by a gradient of transpiration rate, or cuticular permeability. This is illustrated in Table 6.5 and here, for the purposes of comparison, is included the transpiration rate of *Onymacris plana*, a tenebrionid from the Namib that has the lowest transpiration rate so far recorded for one of these desert insects. The rates given in the table span a range that invites comparison with those of the whip scorpion *Mastigoproctus giganteus*, the solifugid *Galeodes arabs*, the tarantula *Eurypelma* sp., and the semimontane scorpion *Diplocentrus*

TABLE 6.5.
Rates of Water Loss in Various Species of Desert Tenebrionids

Species	Water Loss (% body weight hr^{-1})	Authority
Onymacris plana	0.052	Edney 1971
Cryptoglossa verrucosa	0.090	Ahearn 1970a
	0.120	Edney 1971
Centrioptera muricata	0.120	Ahearn 1970a
Eleodes armata	0.200	Ahearn 1970a
	0.245	Edney 1971

Source: Compiled by the author.

spitzeri. However, they are appreciably higher than the rates recorded for such xeric scorpions as *Leiurus quinquestriatus* and *Hadrurus arizonensis* (see Table 6.1).

Desert ants employ a variety of behavioral traits to avoid exposure to drought conditions, as we have already seen. They also rely heavily on the water preformed in their food for their moisture requirements in all probability. Nocturnal foragers may also avail themselves of dew condensates. However, some species of the harvester genus *Pogonomyrmex*, and *P. rugosus* in particular, may be able to produce metabolic water, since Ettershank and Whitford (1973) found that rates of oxygen consumption increased with increases in the vapor pressure deficit. It is perhaps worthy of note, in this context, that *P. rugosus* feeds mainly on seeds and will not accept water even when it is available; there is a parallel here with rodents of the genera *Dipodomys* and *Jaculus* (see below).

Foraging activities of desert ants have usually been considered in the context of the temperature variable. However, desert ants may be subject to desiccation at high vapor pressure deficits, and this may determine the upper limits of foraging activities in some cases. Kay and Whitford (1975), for example, noted that the degree of tolerance to desiccating conditions varied greatly, even within the same genus of ant. Furthermore, harvester species, such as *Pogonomyrmex* spp., forage over the surface of the soil and are more likely to have physiological adaptations to cope with moisture stress than, for example, honey-pot ants (*Myrmecocystus* spp.), which forage in shrub canopies (C. Kay and Whitford 1978). Again, it has been shown that *Formica perpilosa*, which is a nocturnal forager on plant exudates (Whitford, Kay, and Schumacher 1975)

162 Desert Soil Fauna

or insects (Whitford 1978c), has a higher desiccation rate and consequently a lower desiccation resistance than species of *Pogonomyrmex* and *Novomessor*, which forage for seeds during the day. Finally, Délye (1968) reported that desert-dwelling species in the Sahara have less permeable integuments than nondesert species.

Rodents

A relationship can be defined, among surface-active rodents, between feeding habit and the strategies of water gain and water conservation. As we have already noted, these vertebrates are to some extent avoiders by virtue of their nocturnalism and burrowing behavior. They have also developed behavioral, structural, and physiological mechanisms that enable them to tolerate, within limits, the vicissitudes of life in arid environments. At the behavioral level and, more particularly, in the context of feeding behavior, a distinction can be made between "wet" and "dry" rodents. The former, exemplified by species of *Citellus* and *Neotoma*, feed on green vegetation, often supplemented with insect prey, and obtain their moisture requirements from preformed water in their food. The latter, which include the heteromyid *Dipodomys* (Fig. 6.9) and *Jaculus* species, are mainly granivores and do not rely on an exogenous water source. The pocket mouse *Perognathus fallax*, for example, can live indefinitely on dry seeds and selects these in preference to moist food. *Dipodomys* and *Jaculus* species, on the other hand, will feed on green vegetation, particularly during the rainy season, and underground tubers and bulbs at other times of the year (Ghobrial and Nour 1975).

The water relations of desert rodents have been extensively reviewed elsewhere, notably by Chew (1965), Schmidt-Nielsen (1964, 1975), and Ghobrial and Nour (1975), and it would be invidious and indeed unnecessary to dwell on these at great length here. However, there are some points worthy of mention. For example, the utilization of preformed water in the food is not always a simple matter. The pack rat *Neotoma lepida*, for example, obtains much of its water requirements from feeding on cacti, a source that contains high levels of oxalic acid. In the normal course of events this would sequester calcium ions and lead to the impairment of blood clotting. However, *Neotoma* evidently possesses a special physiological mechanism that that prevents this from happening and also allows the excretion of calcium carbonate without the formation of kidney stones (Schmidt-Nielsen 1964). Again, the sand rat of the Sahara, *Psammomys obesus*, is a highly

Fig. 6.9. The kangaroo rat *Dipodomys deserti* from Mojave County, Arizona. (Photo by R. E. MacMillen.)

specialized herbivore (Petter 1975) that feeds on succulent halophilic plants. The sap of these plants may be twice as concentrated as seawater, but *Psammomys* copes with this problem by having a highly efficient kidney that excretes large volumes of urine with very high electrolyte concentrations (Schmidt-Nielsen 1975).

Dry rodents rely on water conservation rather than water gain from their environment. Oxidative metabolism in these granivores yields endogenous water from the seeds on which they feed, and this forms an important part of their water budget. Nevertheless, nocturnal foraging by many of these seed feeders ensures maximum exploitation of whatever water is available in the environment. The seeds on which these granivores feed vary in their water content during the 24-hour cycle. Dew formation is induced by nighttime temperatures in hot deserts, and in some regions this form of precipitation may exceed the total annual rainfall. The condensation of water on seeds during the night, coupled with nocturnal foraging behavior, means that seeds are harvested when their moisture content is at a maximum (Schmidt-Nielsen 1975). Furthermore, the caching of moist seeds in subterranean nests, and the deposition of feces, which subsequently may be refected, in these sites may serve to husband environmental moisture.

The small size—and therefore the high surface area/volume ratio—of desert rodents presents a problem of body water conservation, namely, of potentially high water losses across the skin. This problem is solved, very simply, by the absence of sweat glands, although, as noted earlier, cutaneous water loss appears to be greater in desert rodents than in desert lizards. Adaptions to curtail water loss through respiratory and excretory avenues have also been developed. *Dipodomys merriami*, for example, has a nasal cooling device whereby the surface area of the mucosal membrane is increased, and the air expired from the lungs is cooled as it passes over this mucosa, and much of its moisture condenses here. The production of dry feces and the excretion of extremely concentrated urine are other adaptations shown by desert rodents to conserve body water (Schmidt-Nielsen 1964, 1975; MacMillen and Lee 1967, 1969; MacMillen 1972; MacMillen and Christopher 1975). The production of a concentrated urine is a feature of desert rodents and, although it is developed to varying degrees (MacMillen and Christopher 1975; see also Chapter 8), is achieved by long-looped nephrons in the kidneys that act as a countercurrent multiplier system.

Testudines

Finally, in this section on surface-active primary consumers, attention may be drawn to these herbivores that can obtain water from at least three sources. The succulents on which *Gopherus agassizii* feeds provide a source of preformed water, although this species as well as *Terrapene horsfieldii* evidently imbibe large quantities of free water when this is available (Mayhew 1968). Metabolic water may also be utilized, but this varies with the species since desert tortoises differ in their ability to store fat. *Kinosternon sonoriense* is much more efficient in this respect than *Gopherus agassizii* (see Chapter 3).

Water conservation is practiced in these desert testudines in a number of ways. The shell-like carapace restricts transpiration losses. Schmidt-Nielsen and Bentley (1966) measured the evaporative water loss from the body surface of *Gopherus agassizii* and concluded that this, together with respiratory water loss, was lower than in other chelonians. Further, tortoises retreat into burrows to avoid high temperatures and the need for evaporative cooling mechanisms. However, when such mechanisms have to be invoked

Drought Tolerance   165

they involve the discharge of bladder water over the posterior limbs. Desert testudines can store water in appreciable quantities for relatively long periods of time in the urinary bladder. In such circumstances waste products of nitrogen metabolism will accumulate in the bladder, and their retention here may well be a function of their solubility. Urea, as a form of nitrogenous waste, is considerably more soluble than salts of uric acid. It comes as no surprise, therefore, to find that desert testudines can shift their metabolism from uricotelism to ureotelism. This shift seems to be governed by the relative activities of arginase (which promotes the formation of urea) and xanthine oxidase (which mediates uric acid formation) (Mayhew 1968). Just what determines the balance of this enzymatic activity remains to be disclosed, but the availability of environmental moisture remains a possibility.

Finally, despite their abilities to maximize water gain from the environment, and to minimize body water loss, all desert reptiles may be faced from time to time with situations of water stress. These situations occur particularly when water is lost from the blood to the environment. In the short term this loss may be made good by shunting water from the bladder fluid to the blood, but in the longer term the ability to tolerate large increases in plasma ion concentrations, such as shown by *Gopherus agassizii* (Dantzler and Schmidt-Nielsen 1966), may allow desert reptiles to survive under conditions of considerable dehydration.

THE SUBTERRANEAN FAUNA

It might be supposed that the completely subterranean mode of life adopted by the nematodes and microarthropods discussed in Chapter 4 would isolate them from environmental stress. That this is not entirely true is evidenced by the reproductive strategies employed to ensure maximum survival of the young (Chapter 5). These strategies can be identified as avoidance mechanisms, but evidence is now beginning to accumulate that suggests that some members of this subterranean fauna, at least, have other survival strategies that allow them to tolerate drought conditions in hot deserts. One way in which this tolerance can be achieved is by the assumption of a physiological state known as anhydrobiosis.

The identification of the anhydrobiotic state, and its study, are recent developments in the ecology of desert soil animals. The main reason why this phenomenon has escaped the attention of ecologists until now lies in the simple fact that animals in this state are not metabolically active and cannot be extracted from the soil by conventional behavioral methods (Murphy 1962). Special techniques are required to recover anhydrobiotic invertebrates, and these have been developed up to the present time for only two desert-dwelling groups: the nematodes and collembolans.

Much of our knowledge of anhydrobiosis in desert nematodes comes from the work of Freckman (1978). She noted that anhydrobiotic nematodes assume a tightly coiled spiral and are shrunken, wrinkled, and folded longitudinally and transversely (see Figs. 4.2 and 4.3). This folding reduces the surface area and represents a device for curtailing water loss. Anhydrobiosis in nematodes is not confined to any particular stage in the life cycle, or trophic group, and this state apparently increases the tolerance to temperature extremes and the effects of ionizing radiation. Recovery from the anhydrobiotic state—and the resumption of the normal, vermiform shape—occurs at moisture levels of no more than 12 percent after a 24-hour exposure period in soil. Metabolic activity is resumed prior to this (at 3 percent moisture level). Using these activity levels as a guide, Freckman (1978) calculated that soil nematodes in a desert site at Rock Valley, Nevada, were in an anhydrobiotic state in the field from April to October—that is, during the dry season in this part of the Mojave desert.

Since anhydrobiotic animals are in a quiescent state, they are not participating in community metabolism or the decomposition process. Failure to recognize this fact can lead to overestimates of the extent to which faunal groups, such as soil nematodes, participate in this process. Freckman (1978) took account of the length of the anhydrobiotic period undergone by detritus-feeding nematodes in Rock Valley and arrived at the conclusion that they appropriated about 1 percent of the annual energy input to this desert soil system. However, there are other ways in which they can influence the decomposition process, as already noted in Chapter 4.

The phenomenon of anhydrobiosis, in one form or another, in Collembola has been highlighted by the work of Poinsot (1968, 1974) and Greenslade and Greenslade (1973). These authors noted that certain species belonging, for example, to the genera *Folsomides* and *Brachystomella* respond to low moisture levels in the environment by assuming a wilted, shriveled appearance and, to all intents and purposes, giving the impression of being dead. However, these moribund individuals become reactivated after rain and resume normal mobility. The assumption of this quiescent state is only one of a suite of adaptations shown by Collembola in arid environments (P. Greenslade, in press). Members of the genera *Xenylla*, *Prorastriopes*, and *Corynephoria*, for example, exhibit a physiological tolerance that allows them to remain active during drought conditions. *Brachystomella*, *Setanodosa*, and *Folsomides* species produce drought-resistant eggs that hatch in the presence of free water. Unfortunately, little is known at the present about the mechanisms underlying these various expressions of physiological tolerance. It may be worthy of note, however, that species of the genera *Prorastriopes*, *Corynephoria*, and *Sphaeridia* belong to the Sminthuridae. Members of this family have also been recovered in large numbers from mesquite litter in the Chihuahuan desert (Wallwork, in preparation), and evidently they flourish in arid sites. These Collembola have a tracheal method of respiration; as a consequence the general body surface does not function as a respiratory surface (as it does in many other collembolans). This means that the sminthurid cuticle can be made waterproof with an epicuticular lipid layer. The possession of this water barrier will undoubtedly restrict transpiration loss, although this remains to be quantified. It allows sminthurids in general to colonize surface litter and epigeal habitats, and they are obviously preadapted to life in dry environments.

The ability to pass into an anhydrobiotic state is in itself an interesting adaptation to life in a xeric environment. It is rendered even more intriguing by the fact that this phenomenon has developed independently in nematodes and Collembola—two groups that are phylogenetically remote from each other. Of course it is conceivable, indeed likely, that anhydrobiosis in nematodes is of a fundamentally different nature from the inactive state occurring in Collembola (Diana Freckman 1979: personal communication). However, the question naturally arises as to how widespread may be the wilting phenomenon among desert soil invertebrates. It is well known, for example, that epigaeic oribatid mites that burrow into the thallus of lichens become inactive during dry periods and are roused when wetted (Travé 1963). Similar behavior patterns are shown by other mites of this group, such as *Alaskozetes antarcticus* of cold polar deserts. However, it is unlikely that many of the adult oribatids that live in hot desert soils undergo anhydrobiosis in the same way that nematodes and Collembola do since the shell-like, rigid exoskeleton of these mites will not allow the body to wilt, or become wrinkled and shriveled. They must possess some other physiological tolerance mechanism that ensures survival during periods of drought—and this remains to be elucidated. On the other hand, the juveniles of this group of mites are soft bodied, and the flexible cuticle can be folded to restrict water loss. Indeed, some are preadapted in this respect; parenthetically, the juveniles of *Alaskozetes antarcticus* have a strongly pleated cuticle that may function as a means of water conservation in cold arid environments.

Likewise, the immatures of *Tectocepheus* and *Passalozetes* species, which have been recovered consistently from soils in the Chihuahuan desert (Wallwork, in preparation), have folded cuticles. Neither of these two genera is restricted to hot deserts, but their predilection for dry environments, particularly in the case of *Passalozetes*, gives them a selective advantage in exploiting such habitats. Another group of mites, the Prostigmata, which flourish in hot desert soils, are soft bodied both as juveniles and adults and have the potential for assuming the anhydrobiotic state. Likewise the Pauropoda (see Chapter 4), which are locally abundant in the organic soils of hot deserts. Virtually nothing is known about the survival strategies of these arthropods, however, and this is an area that lends itself readily to further research.

Earlier in this book it was argued that rainfall is the driving variable in hot deserts—that biological activity ceases between rainfall events and that new pulses of activity are triggered by new inputs of precipitation. In a broad sense these statements may well be true, but they require some qualification. It is true, for example, that major inputs of moisture to the desert soil system are discontinuous in nature, and the same is probably also true for the input of organic material. In the previous pages we have also reviewed the considerable body of evidence that indicates that reproductive activity (a major "growth" compartment) is often opportunistically geared to these environmental pulses. Locomotory activity may be similarly restricted particularly, but not exclusively, in seasonally surface-active animals such as millipedes, snails, tenebrionid beetles, ants, tortoises, amphibians, and rodents; the truly subterranean nematodes and Collembola may exhibit dormancy during the dry season, as just indicated. However, dormancy is not a universal feature of desert soil animals; it does not occur in surface-active predatory invertebrates such as the centipedes, scorpions, solifugids, and uropygids, nor in some groups of subterranean mites. Studies in the Mojave (Wallwork 1972a) indicated that the cryptostigmatid mite *Joshuella striata* and the prostigmatid *Spinibdella cronini* were active during extended periods of the year. The extent of this surface and subterranean activity could be a function of the environmental temperature. This provides a convenient note on which to end this chapter and start the next.

7
The Temperature Factor and the Biological Response

The twin effects of moisture and temperature are notoriously difficult to separate in the soil ecosystem, and, as Noy-Meir (1973) points out, temperature effects interact closely with the water factor in hot deserts. In the preceding pages we have already experienced difficulty in this regard, particularly when considering the deployment of precious body water for thermoregulatory purposes. On the other hand, hot desert systems are probably more amenable than most to the separate study of moisture and temperature effects, if only because the moisture factor at least can be treated as a discontinuous variable. Pursuing this line of reasoning a little further brings us to the point where we can identify stochastic patterns of rainfall as providing an overall trigger for a major biological response. However, the intensity of this response, generated from some reserve compartment, is probably a function of temperature, at least to some extent. Looking at this in a slightly different way, it is possible to view the moisture factor as an overall governor of the system, providing the coarse adjustment that sets a train of biological events in motion. The precise character of this response may be more finely tuned by the temperature factor. The interaction of moisture and temperature effects can, in this way, be divided into responses operating at two separate levels: coarse tuning and fine tuning, respectively. This view of desert ecosystem functioning may have a generality that is worth investigating. However, it may not have universality if note is to be taken of Kay and Whitford's (1975) finding that foraging activities of Chihuahuan desert ants may be limited, in the last analysis, by vapor pressure deficits. This does not invalidate the premise, though, since desert ants are well able to cope with the environmental temperature factor, by virtue of their mobility if for no other reason. Many other desert soil animals are less well equipped in this regard, and it is these with which we are particularly concerned here. Their response to the temperature factor is important. But before we consider this in more detail, the ways in which this factor is expressed in hot desert soils must be examined.

It is convenient to analyze the temperature factor in two dimensions: spatial and temporal. Each of these can be subdivided further. The spatial dimension has global and local components; the temporal dimension has seasonal and diurnal components.

GLOBAL PATTERNS

In Chapter 1 a distinction was made between hot and cool deserts; this can now be amplified. Viewed in the context of their temperature regimes, the major desert areas of the world are complex and lack uniformity. There are several factors that contribute to this heterogeneity, and in surveying these briefly we can turn to the classification proposed by Walter and Stadelmann (1974) and the reviews of McGinnies (1979a, 1979b). In the first instance there are latitudinal variations in temperature. Using this criterion, for example, we can distinguish the low latitude deserts of the tropics from the Mediterranean-type deserts. To the former belong the Sahara, Arabian, and Indian deserts, which are characterized by large annual variations in temperature. Here, midday temperatures may reach 40°C to 45°C during the high-sun period, while minimum night temperatures during the period of low sun average about 10°C. Even so, winters tend to be warm in these deserts, with daily maxima in the range of 15°C to 25°C. In contrast the Mediterranean-type deserts have mild or cool winters. These are mainly located in more northerly latitudes and are exemplified by the northern fringes of the Saharan and Arabian deserts, the Israeli Negev, the Iranian desert, and parts of the North American deserts. The southern region of the Australian desert also has a Mediterranean climate. At even higher latitudes warm or hot summers alternate with subfreezing winters to characterize what we are here considering to be cool deserts, such as the Takla Makan/Gobi complex of central Asia, the Great Basin desert of North America, and the Patagonian desert. In the Takla Makan, high altitude is also a factor that contributes to low temperatures in winter. Much of this desert is highly mountainous as is the Monte-Patagonian desert. The climate of the latter is also influenced by winds coming from the east off the cold Falkland current; as a consequence the summers are mild and the winters cool. The temperature regime of such coastal deserts as the Namib and the Atacama is determined at least in part by their proximity to cold ocean currents, and here we have yet another factor that has a bearing on the temperature characteristics of desert regions.

LOCAL PATTERNS

It will be evident that in a general sense the temperature regimes of the world's deserts are determined by geographical factors: latitude, altitude, and proximity to coastal regions. However, within a given desert, even within a small area of desert, temperatures are not uniform at any particular point in time. Local variations occur, very often as a function of patterns of sun and shade; because the soil surface represents an interface between the soil itself and the air above, these variations will be at a maximum here. This is illustrated by the temperature profiles shown in Fig. 7.1, which were measured at a site in the Mojave desert. These profiles demonstrate the considerable protection from intolerable surface temperatures that the shade of vegetation provides for animals living in the top few centimeters of the soil. It is these temperature conditions that constitute part of the immediate environment of the soil fauna and to which this fauna responds directly. However, it is also evident from the figure that the amplitude of the temperature wave is considerably dampened at a depth of a few centimeters, even in unshaded conditions, and this attests to the efficiency of the soil as a heat barrier. Exposed soils in hot deserts present no formidable obstacles for subterranean animals as far as their temperature regimes are concerned. However, other factors such as moisture and food availability may prove to be limiting here. Perhaps it is no coincidence that the most successful colonists of such soils are animals that import their food supplies from above ground, such as the harvester ants and granivorous rodents.

Another factor that affects patterns of temperature is aspect. This is likely to find its greatest expression where the action of wind and water has produced a dissected topography: slopes and slip faces, dune crests, ridges and bases. In the Namib desert, for example, dune topography is an important determinant of species' distribution among Tenebrionidae (see Chapter 8). In the Negev

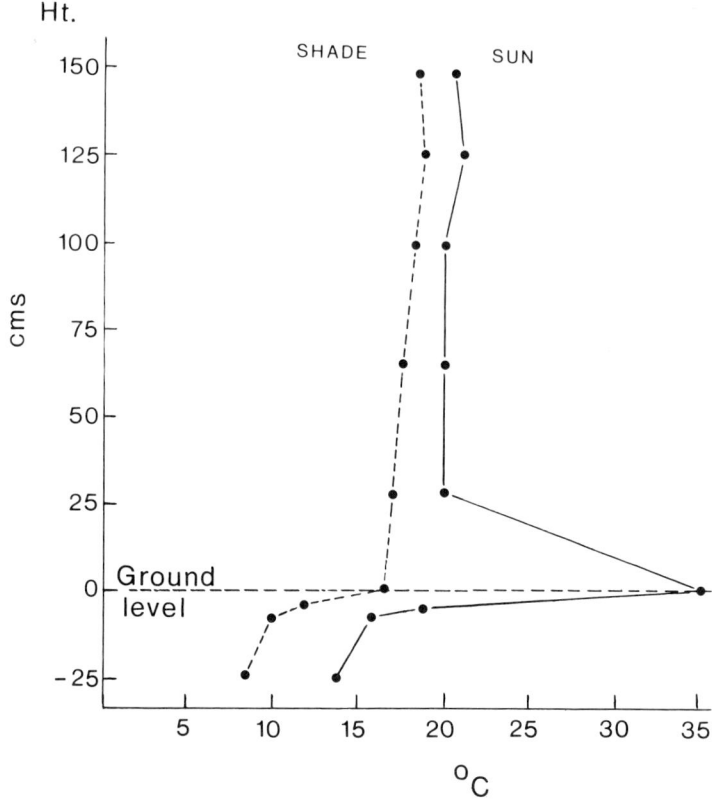

Fig. 7.1. Temperature profiles recorded at 1100 hours during November from a site shaded by juniper and an exposed mineral site nearby, in the southern Mojave desert. (From Wallwork 1972a.)

desert north-facing slopes provide less inhospitable temperature regimes than south-facing ones, and this is reflected in the distribution of snails of the genus *Cristataria* (Bar 1977; see also Chapter 9) and the woodlouse *Hemilepistus reaumuri* (Brown, unpubl.).

SEASONAL PATTERNS

That air, and therefore soil, temperatures vary from season to season can be taken as axiomatic. However, the less obvious fact is worth stating—namely, that amplitude of the seasonal temperature wave varies considerably from one desert to another. As already noted earlier in this chapter, low latitude deserts are characterized by seasonal variations on the order of 15°C to 25°C; these are the deserts with hot summers and warm winters. They contrast with the cool deserts, such as the Gobi, which may have hot summers and cold winters. Here, temperatures can vary from −25°C in winter to +44°C in summer (McGinnies 1979b). A further contrast is provided by the Atacama desert of Chile and Peru in which there is very little temperature variation between summer and winter, and a generally mild climate prevails.

Seasonal temperature patterns in deserts are intrinsically interesting because of the direct influence of temperature on biological activity; sadly, the variations that can be played on this theme have received little attention, although Wallwork (1980) has attempted to identify some of them (see also later in this chapter). However, it is even more significant, from the biological point of view, to examine the temperature factor in relation to the seasonality of rainfall. Noy-Meir (1973) has noted that this relationship has a strong modifying effect on plant growth dynamics, and the possibility that this idea can be extended to soil faunal activity cannot be overlooked. In low latitude and summer rainfall deserts, for example, soil moisture and temperature will be simultaneously optimal, and the biological response is immediate and rapid. Such a situation will obtain in the Sahara, the northern part of the Great Australian desert, and the Chihuahuan desert of the southwestern United States. At the other extreme are the winter rainfall deserts of Central Asia, the Mojave desert of California, and the Great Basin desert of Nevada. Here, although moisture is plentiful during the winter, low temperatures at this time may inhibit some forms of biological activity, and this activity is only resumed when temperatures ameliorate during the following spring.

DIURNAL PATTERNS

Superimposed on the seasonal pattern of temperature fluctuation is a circadian pattern that can have a direct effect on the locomotory and foraging activity of soil animals in deserts. Crawford and Riddle (1974) monitored temperature changes at the soil surface and at various depths in the soil at sites in the Peloncillo Mountains of southern New Mexico over 24-hour periods during the fall and also during the winter. Circadian profiles typical of the fall season are given in Fig. 7.2 for temperatures at the surface and at a depth of 8 cm. These demonstrate quite clearly the buffering effect produced by the top few centimeters of soil. This effect persists into the winter when surface temperatures may fall below zero, particu-

174   Desert Soil Fauna

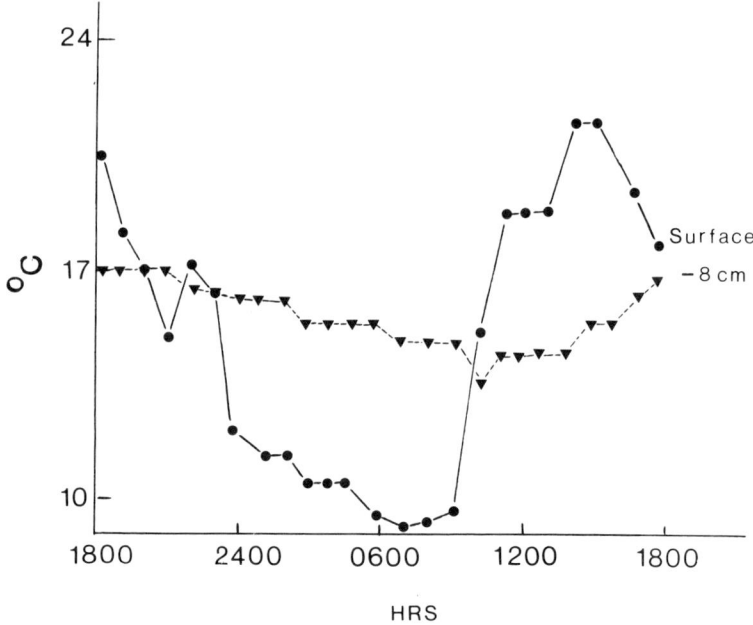

Fig. 7.2. Variation in temperature at the soil surface and at a depth of 8 cm in October at a mountain site in the Peloncillo Mountains, New Mexico. (After Crawford and Riddle 1974.)

larly in montane areas. At this time of the year, the soil may experience zero temperatures down to a depth of 8 cm; clearly, animals that produce only shallow burrows, such as the centipede *Scolopendra polymorpha*, the scorpion *Diplocentrus spitzeri* (see Crawford and Riddle 1974), and desert snails, are at risk at this time of year.

The direct effects of temperature at the soil surface may limit locomotory and foraging activity either because the temperatures are too high or too low. In addition to these direct effects there may be more subtle influences within the soil profile itself. Once again, these are a function of the interrelationship between temperature and moisture. As Noy-Meir (1973) has pointed out, moisture vapor can move along soil temperature gradients in dry soils subjected to high daily radiation. This movement can cause internal dew formation as a result of the upward flow of moisture, and this condensation effect may stimulate biological activity during the early hours of the day, even during the dry season. Whitford, Freckman, Elkins, Parker, Parmalee, Phillips, and Tucker (in press) and Whitford, Meentemeyer, Seastedt, Crumack, Crossley, Santos, Todd, and Waide (1981) have noted that oribatid mites have such activity windows in the Chihuahuan desert. This observation merits further study since it suggests that biological activity in this system is not as discontinuous as the Noy-Meir (1973) model postulates (see Chapter 5).

THE BIOLOGICAL RESPONSE

In previous chapters we have reviewed the various ways in which soil animals in arid regions avoid or tolerate drought conditions. Avoidance is very much a matter of developing behavioral (including reproductive) patterns in which the ability to burrow, nocturnalism, and seasonality combine to remove animals from situations of environmental stress. Tolerance mechanisms, on the other hand, are largely physiological in nature, and these place emphasis on the ability to conserve water once it has been taken into the body. Such mechanisms allow animals to endure periods of environmental moisture stress.

These twin ideas of avoidance and tolerance can be applied with equal force in interpreting the biological response to the temperature factor in hot and cool deserts. In winter rainfall deserts, summer drought is associated with high ambient temperatures, and this combination of moisture and temperature variables may result in a concentration of biological activity in the spring months—when the effects of winter rains may still linger and temperatures have not yet reached their summer maxima. However, by focusing down—in a temporal sense—from the seasonal to the circadian level, the response to temperature becomes more sharply delineated from the moisture response. This brings in a dimension that is sometimes overlooked in consideration of hot desert ecosystems, namely, that animals that inhabit such systems have to cope not only with high environmental temperatures but also, on occasion, with zero or subzero temperatures (see earlier). Freezing temperatures are common during the winter months in cool deserts and, indeed, in hot deserts. The ability to withstand these cold periods may be at least as important as the ability to withstand high temperatures, as far as desert animals are concerned. Bearing these points in mind, we can now proceed to examine the ways in which desert soil animals come to terms with the temperature factor.

Avoidance

The faunal groups that lead a completely subterranean existence would seem not to require, at first sight, any special mechanisms for avoiding temperature stress because of the buffering effects already noted (Figs. 7.1 and 7.2). However, the adoption of a completely subterranean mode of life requires adaptations, and groups that have assumed this life-style must be able either to burrow efficiently or to move through the cavities that exist between soil particles. The adaptations associated with burrowing have already been dealt with at some length in Chapter 5 and need not be repeated. On the other hand, the interstitial fauna can only penetrate the soil to a depth that is determined by their body size and the cavity structure of the soil. This depth is rarely more than a few centimeters, and the faunal groups so limited in their distribution, which comprise the majority of soil microarthropods in hot deserts, may be exposed to some environmental temperature fluctuation diurnally and seasonally, particularly in relatively unshaded sites.

That this exposure indeed occurs, and produces a faunal response, is indicated by an example from the Mojave desert that was introduced in Chapter 4. This example is drawn from a study of the microarthropod fauna in a juniper litter site (Wallwork 1972a) that experiences winter rainfall (December) when the average daily maximum temperature is about 16°C. One of the characteristic species in this site is the oribatid mite *Haplochthonius variabilis*, which evidently produces a new generation not immediately after the rains when temperatures are low but three to four months later (see Fig. 4.12) when average daily temperatures are substantially higher.

There are some circumstantial reasons then for supposing that *H. variabilis* has devised a tactic for avoiding the environmental stress occasioned by low temperatures: a tactic that involves delaying its period of recruitment until temperature conditions become favorable. This supposition requires empirical confirmation. Further, this tactic may not be a general phenomenon. In this same Mojave juniper site another oribatid mite, *Joshuella striata*, responds immediately to December rains by recruiting at this time, despite the relatively low temperatures. This suggests that there may be interspecific differences with regard to the response to temperature (see below).

Avoidance often demands a degree of flexibility in behavior pattern. In documenting this statement we must anticipate a little of what is to follow, particularly with regard to the ants.

For surface-active species, such as ants, the question is not so much a matter of timing reproductive activity to coincide with favorable temperature regimes as it is a matter of managing the more immediate problem of maximizing foraging times on the surface. Such foraging activity has to occur within the thermal limits of the ants, but any relaxation of these limits will allow for an extension of the period when ants will be able to acquire food on the surface. Such adaptations to the thermal environment are really tolerance, rather than avoidance, mechanisms and will be considered in this context in a moment. They are mentioned at this point to emphasize the distinction that may be made between them and the behavioral adaptations that involve a temporal shift of the activity period during the 24-hour cycle. Such shifts occur as a response to changing seasonal conditions. Whitford and Ettershank (1975) have shown, for example, that *Pogonomyrmex* spp. and *Novomessor cockerelli* can shift from diurnal to nocturnal foraging activity during the summer months in the deserts of the southwestern United States. This may be a device to exploit a seasonal abundance of nocturnal prey in the case of *Pogonomyrmex* spp., which will feed on termites as an alternative to seeds. On the other hand, it may be a reaction to avoid high daytime temperatures in the more "generalist" feeder *N. cockerlli.*

Other examples can be quoted to show a shift in the opposite direction during winter. As Ghosh (1975) has pointed out, dawn and dusk temperatures may drop to unacceptable levels during winter, and surface-active rodents such as *Meriones hurrianae*, which are crepuscular during the summer, shift their activity to later in the day during winter. When daytime temperatures during the winter become intolerably low, surface activity ceases; the wood louse *Hemilepistus reaumuri* illustrates this very nicely.

In the Negev desert *H. reaumuri* exhibits a pattern of diel activity that varies with the seasonally related phases in its life cycle (Shachak 1980). This isopod has five distinct phases in its life cycle (see Chapter 2), and the patterns of surface activity associated with each of these phases are shown in Table 7.1. These data show that surface activity is bimodal during the period of gestation and growth (April to November) and unimodal during the time of pair formation (February to March) but that it ceases completely during the period November to January. This cessation of surface activity coincides with the period of lowest burrow temperatures, although it is not known if *H. reaumuri* becomes dormant at this time or merely remains active within the burrow.

Although much has already been said about burrows as refugia for soil animals, we still need to consider their temperature

TABLE 7.1.
Patterns of Diel Activity in Relation to Season and Life Cycle Phases in the Desert Woodlouse *Hemilepistus reaumuri*

Life Cycle Phase	Month	Hours of Activity	Burrow Temperature (°C)
Pair formation	February	1200-1600	11-17
	April		16-34
Gestation		0700-0900	
Hatching	May	1700-1800	16-25
Growth	November	0500-0700	15-24
Stationary	January	–	7-13

Source: Based on Shachak 1980.

characteristics. These characteristics will depend on the type of refugium—deep burrow, shallow burrow, vertical or horizontal shaft, crevice or shelter under rocks or vegetation. The amount of protection provided by these refuges varies. Ghosh (1975) and Ghobrial and Nour (1975) showed that temperatures in the burrows of desert rodents rarely exceed 33°C (confirming the findings of Misonne 1959) even when soil surface temperatures are in excess of 70°C and that temperature variations in the burrows of *Meriones hurrianae* span no more than 5°C over the annual cycle. The Sudanese jerboa, *Jaculus jaculus*, ensures an equable environmental temperature regime by constructing different kinds of burrows at different times of the year (Ghobrial and Nour 1975). During the summer and winter, when surface temperatures are likely to be extreme, it produces a sharply inclined burrow system leading to a nest compartment that may be as much as 80 cm below the surface. In autumn, however, the burrows are shallow and almost horizontal with the nest compartment situated at a depth of no more than 20 cm.

Temperature characteristics in and around the surface nests of the North American pack rat, *Neotoma lepida*, have been described by Schmidt-Nielsen (1964), and these are illustrated in Fig. 7.3. The nest itself is constructed from twigs, branches (see Fig. 5.9), or fragments of cacti (Brown, Lieberman, and Deugler 1972). The temperature in the interior of the nest does not usually fluctuate beyond the tolerance limits of these rodents, and this refuge provides a means of avoiding high ambient temperatures found elsewhere at the soil surface during the summer.

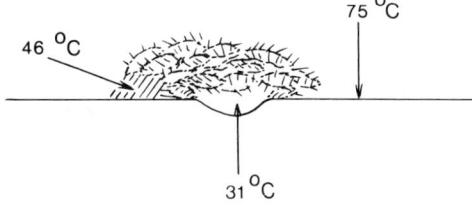

Fig. 7.3. Temperature characteristics in and around the nest of the American pack rat, *Neotoma lepida* (After Schmidt-Nielsen 1964.).

Animals that seek the shelter of preexisting crevices, rather than excavated burrows, are at more of a disadvantage in deserts, if only because they place themselves at the mercy of their environment. A good example of this is provided by the centipede *Scolopendra polymorpha* in the Chihuahuan desert of New Mexico. This desert, like many other hot deserts, experiences subzero temperatures at the soil surface during winter (see earlier). Animals living at or near this surface must be able to avoid or tolerate these low temperatures. *S. polymorpha* does not appear to have the ability to withstand cold to any extent (Cloudsley-Thompson and Crawford 1970; Crawford and Riddle 1974; Crawford, Riddle, and Pugach 1975). It must either withdraw into deep soil burrows or perish. Crawford and Riddle (1974) collected *S. polymorpha* during the winter from surface microhabitats in New Mexico; these specimens were practically immobilized by cold and clearly had not sought refuge deeper in the soil. These workers estimated that 10 percent of each breeding population of this centipede in the Chihuahuan desert suffered mortality due to cold. Mortalities in excess of this are probably sustained by the scorpion *Diplocentrus spitzeri*, which occurs sympatrically with *S. polymorpha*. This shallow-burrowing scorpion retires beneath rocks (Crawford and Krehoff 1975) and is no more cold adapted than *S. polymorpha* (Crawford and Riddle 1974).

The soil is clearly the main refuge from temperature extremes for animals that must be active on its surface from time to time. But it is not the only avenue of escape; aerial vegetation provides another—and here again, temperatures are lower than at the soil surface (Fig. 7.1). The iguana of North American deserts, *Dipsosaurus dorsalis*, is a surface forager that retreats into the shade of the creosote bush during the hottest part of the day. Likewise, the horned lizard *Phrynosoma cornutum* (see Fig. 3.3) spends a considerable proportion of its diel activity in shrub climbing and movement within the shrub canopy. Whitford and Bryant (1979) observed this behavior in the canopies of *Yucca elata* and *Ephedra trifurca* and noted that the lizard could maintain its body temperature between 35°C and 40°C in these situations. In Old World deserts the palm squirrel *Funambulus pennanti* shows a similar escape behavior (Ghosh 1975). *D. dorsalis, P. cornutum,* and *F. pennanti* cannot really be considered as committed members of the soil fauna, however, even though they are active on the soil surface during daylight. Collembolans of the genera *Deuterosminthurus* and *Corynephoria* have closer links with the soil fauna, and members of these groups evidently migrate upward into the aerial vegetation when air temperatures exceed 20°C (in the case of *Corynephoria*) or 30°C (in the case of *Deuterosminthurus*) (P. Greenslade, in press). Similarly, certain honey-pot ants, notably *Myrmecocystus romainei, M. depilis,* and *M. mimicus,* show extended periods of daytime activity. They do this by moving into shrub canopies during the hottest part of the day (C. Kay and Whitford 1978). This behavior allows them to forage for nectar when few other ant species are active—and this in spite of their inability to endure unusually high temperatures. The desert millipede *Orthoporus ornatus* provides us with yet another example of this behavioral tactic.

As already noted in Chapter 2, *Orthoporus ornatus* forages on the soil surface mainly during the day in the rainy season. Wooten, Crawford, and Riddle (1975) showed that at this time of year feeding activity ceased when temperatures in the immediate vicinity of the millipede approached 35°C and only resumed when temperatures dipped below this point (Fig. 7.4). As a consequence locomotor and feeding activity were restricted to early morning and late afternoon. At other times of the day *O. ornatus* assumed a coiled or partly coiled state in the shaded parts of aerial vegetation or, if this vegetation was not available, in mammal burrows or crevices beneath volcanic rocks. These authors also showed that *O. ornatus* is opportunistic in its response to the temperature factor. It moves laterally within the shrub canopy to find the most favorable microsites; if shrub canopy temperatures exceed the tolerable limit, it will descend to subsurface refuges. Again, on overcast days activity above ground will extend into the midday period; on occasion nocturnal feeding activity was noted. This millipede is an example of a member of the desert soil fauna that can only exploit its food sources for a limited period of the year. The extent to which it can do this successfully is determined in large measure by its ability to avoid the inhibiting effects of high temperatures. But it also has a physiological trick up its sleeve, as we will see a little later.

Poikilothermous animals in hot deserts are very much at the mercy of the vagaries of environmental temperature fluctuations, and patterns of behavior that permit minimum exposure to tem-

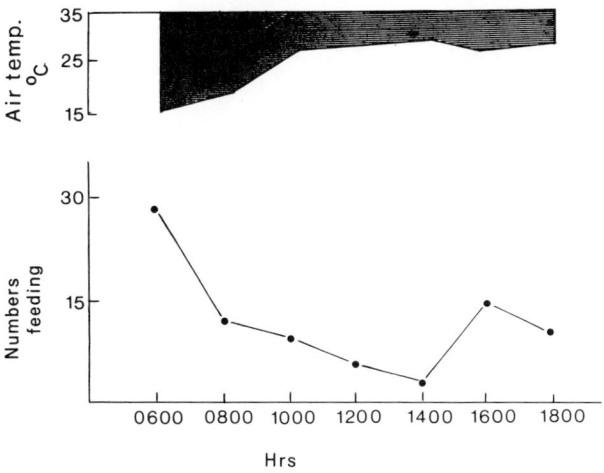

Fig. 7.4. Activity of *Orthoporus ornatus* in relation to temperature. (From Wooten, Crawford, and Riddle 1975.)

perature stress are a vital part of the life-style. The orientation of *Orthoporus ornatus* within the foliage of one of its preferred food plants, Mormon tea (*Ephedra* sp.), is a case in point. Another is provided by the North American desert scorpion *Hadrurus arizonensis*. This nocturnal predator retires during the day to its burrow in the soil and, unlike *O. ornatus*, does not exploit the arboreal avenue of escape. However, it varies its position in its subterranean burrow, moving up and down as the conditions demand to avoid exposure to temperature stress (Hadley 1970b).

These behavior patterns, particularly the seasonal ones, are reinforced by physiological adaptations, such as estivation and hibernation, which have elements of both avoidance and tolerance. As such these adaptations form a convenient link between this section and the next. We have already noted that many desert soil animals become inactive in their refugia at times of the year when surface conditions preclude foraging activity. These include the ants, tenebrionid beetles, and rodents, among others, and their inactive state is characterized by a lowering of the metabolic rate, thereby reducing energy expenditure to a minimum. Another expression of this inactivity is found in the "wilting" of Collembola and assumption of the anhydrobiotic state in nematodes already referred to. This quiescent state may be associated with a relaxation of upper critical thermal limits, which implies some degree of physiological tolerance. Again, an extreme case of estivation is shown by desert snails, such as *Sphincterochila zonata* and *Buli*-

mulus dealbatus, which can remain inactive on the desert soil surface for many months. Schmidt-Nielsen, Taylor, and Shkolnik (1972) have shown that the metabolic rate of dormant *S. zonata* is so low that the snail can survive for several years in this condition.

Tolerance

It could be argued, then, that estivation in the case of *Sphincterochila zonata* is one of a suite of characters possessed by this snail to tolerate, rather than avoid, drought and temperature extremes at the soil surface. The features of this physiological state are probably similar in essentials to estivation in the millipede *Orthoporus ornatus* in which it is more closely associated with an avoidance tactic (withdrawal to soil refugia). This illustrates the difficulty in classifying physiological adaptations into one or other of these two compartments, and perhaps there is little virtue in attempting to do so. However, mechanisms for temperature tolerance are found not only in *S. zonata* but also in a wide range of desert soil animals, many of which can also be considered as avoiders. Some measure of tolerance is required by these animals if they are, for example, to maximize their periods of foraging activity on the soil surface. It may also be necessary for such surface foragers to withstand short exposures to high temperatures occasioned by sudden changes in cloud cover. Some of the ways in which this tolerance is achieved can now be reviewed.

Size and Color. Many desert animals are larger in body size than their mesic relatives. Above ground the jackrabbit and desert fox may be cited as examples. More closely associated with the soil are the large millipedes, scolopendromorph centipedes, tarantula spiders, and tenebrionid beetles. The low surface area/volume ratio of these relatively large-sized forms serves to minimize the effects of high ambient temperatures by exposing relatively little of the body surface to the sun's rays. In tenebrionid beetles the large body size also allows the development of a large air-filled subelytral space, which acts as an insulator against high ambient temperatures (Hadley 1970a). In this regard the ability of the elytra to reflect heat, rather than absorb it, deserves attention.

Most tenebrionids are black in color, and the survival value of this kind of coloration is still the subject of some controversial debate among desert entomologists. Some forms have white elytra, however, and it might be supposed that these would experience less thermal stress than their uniformly black relatives, if only because black surfaces absorb solar radiation whereas white surfaces reflect it. Indeed, Hadley (1970a) showed that subelytral temperatures were lower in white forms, compared with black, but that the difference was not marked. Evidently, coloration does not figure prominently in the survival of these desert insects. To understand why this is so we must digress briefly to consider a parallel case, a vertebrate that has had some limited success in colonizing hot deserts: man himself.

In these deserts man exists in two color varieties as far as his clothing is concerned: white and black. White clothing is a feature of many desert Bedouin, although there are some who wear black robes. Very recently, a study was conducted by Shkolnik et al. (1980) to examine the effect of cloth color on heat gain by a human subject in the Negev desert. Briefly, the results of this study indicated that black robes gain two to three times more heat from solar radiation than do white robes, but the skin temperature of the subject was the same (33°C) in both cases. This was attributed to the fact that black robes with their higher temperature induce a bellows or chimney effect that enhances convection and conveys heat away from the skin. This mechanism can only operate if the robes are loosely worn so that a cushion of air is present between the robes and the skin and there is an adequate access for air under the skirt of the robes. Desert tenebrionids may operate in much the same way. They have a cushion of air (the subelytral space) that separates the surface of the body from the external environment, and the fact that subelytral temperatures were not markedly different in white and black forms suggests enhanced cooling in the latter, possibly a chimney effect, although this remains to be tested. In any event it does not seem to matter very much whether a desert tenebrionid is black or white.

The idea of the insulating properties of an air cushion has been adopted independently by the desert snail *Sphincterochila zonata*. The shell of this snail has a conspicuously large lower whorl that is filled with air. Unlike many other species *S. zonata* does not seal to the substratum, but it withdraws into the upper chamber of the shell when fully exposed to the sun. The air cushion that forms below it then ameliorates the effects of high temperatures generated in the rock to which it is attached. The shell of *S. zonata* is white in color, and it is relatively thick—features that, again, will reduce thermal stress. Indeed, this shell reflects over 90 percent of the radiant energy falling on it (Schmidt-Nielsen, Taylor, and Shkolnik 1972). Further, when this snail withdraws into the upper portion of its shell, it assumes a dormant state (see above). One of the features of this dormancy is the development of a relatively high lethal temperature.

Lethal Temperatures. Poikilotherms living in hot deserts have to face the problem of overheating. They do this in several ways, one of which is to develop a tolerance to relatively high body temperatures. Schmidt-Nielsen, Taylor, and Shkolnik (1972), for example, found that the upper critical limit for *Sphincterochila zonata* varied in the range of 50°C to 55°C (depending on the duration of exposure). This is relatively high for a mollusk, indeed for any invertebrate, except for certain ants (see below), and it reflects the fact that *S. zonata* normally lives in exposed situations. The sympatric *Trochoidea seetzeni* has an even higher lethal temperature, and this may be attributed to the fact that the shell of this species is more flattened than that of *S. zonata*, and the air cushion in the large whorl provides less insulation.

The possibility has already been mooted, earlier in this chapter, that certain ant species may show a relaxation of critical thermal limits and may thereby prolong surface activity. This relaxation can occur at the upper or at the lower thermal limit. In the former case it occurs in day-active species; in the latter, in nocturnal foragers.

Whitford et al. (1981) have reported that three species of harvester ants belonging to the genus *Pheidole* were active at soil temperatures between 15°C and 35°C in the Chihuahuan desert. Similar observations were made on harvesters in the desert of Baja California, Mexico (Wallwork 1976). All of these ants showed peak activity at sunrise, and this activity ceased about midmorning. This pattern of activity is typical of many species of desert ant, and it indicates that in general these arthropods do not have unusually high upper critical thermal limits. However, harvesters belonging to the genus *Pogonomyrmex* and *Novomessor cockerelli* can remain active on the soil surface at temperatures up to 52°C (Whitford and Ettershank 1975) (Fig. 7.5), that is, far into the day. This adaptation may be related to the fact that these harvesters forage on the soil surface rather than in shrub canopies as do, for example, *Myrmecocystus* spp. and *Formica perpilosa* (Schumacher and Whitford 1974; Kay and Whitford 1978).

Pogonomyrmex californicus, P. rugosus, and *Novomessor cockerelli*, which are strictly or facultatively day-active, can remain surface-active down to 4°C or 5°C. This suggests an extended period of seasonal activity, when considered in conjunction with their high upper critical thermal limits. Relaxation of lower critical thermal limits is also likely to be of benefit to more strictly nocturnal foragers. It comes as no surprise to note, therefore, that the night-active honey-pot ant *Myrmecocystus mexicanus* remains active at much lower temperatures than day-active members of this genus (C. Kay and Whitford 1978).

Fig. 7.5. Nest entrance of *Pogonomyrmex californicus*, Chihuahuan desert. Workers are active on the surface of the mound. (Photo by W. G. Whitford.)

Woodlice from arid and semiarid habitats show analogous patterns. The desertic *Venezillo arizonicus* has an upper critical limit of 42°C, comparable with that of the semixeric *Periscyphis jannonei* and *Metaponorthus pruinosus* (41.5°C: Edney 1968b). In contrast the upper critical thermal limit for the mesic *Oniscus asellus* is around 34°C under comparable conditions. Edney (1964a) was also able to show that lethal temperatures varied in different populations of the same species as a function of the environmental temperature. Working with samples of *Armadillidium vulgare*, *Porcellio laevis*, and *P. scaber* collected from sites ranging from California to British Columbia, he demonstrated that the upper critical thermal limits were environmentally determined since samples from each geographic locality, conditioned at 20°C for two weeks, all showed the same lethal temperature.

The ability to tolerate high body temperatures confers several advantages on a desert animal. First, it can store heat in the body during the hottest part of the day and dissipate this heat by radiation and convection during the night. This process avoids the costly expenditure of water that would have to be used for evaporative cooling. Second, it allows for an extension of the periods of surface foraging activity. Third, it gives flexibility to this activity such that it can be adjusted seasonally.

As noted earlier, evaporative cooling is not in general use as a thermoregulatory device by desert animals, although it may be resorted to occasionally during short periods of acute thermal stress. Desert tortoises, for example, will urinate over the hind limbs; desert rodents will salivate on the fur covering the throat region to alleviate this kind of distress. These rodents are, of course, homiotherms, and the necessity to maintain a constant body temperature brings with it a special set of problems.

Some desert homiotherms, such as the camel, can store heat in the body by tolerating temperature fluctuations of 5°C to 6°C (thus becoming effectively poikilothermous), and they can also employ sweating as a cooling device. Desert rodents have no sweat glands, however, so they have no recourse to this method of lowering the body temperature. According to Ghobrial and Nour (1975), their lethal temperatures are not significantly higher than those of nondesert species, although a perusal of the literature indicates that the thermal relations of these desert rodents cannot be dismissed so lightly. Many workers, including Bartholomew and MacMillen (1961), suggest that these desert homiotherms withstand high ambient temperatures by a tolerance of increased body temperature rather than by better thermoregulation. This view receives support from the work of Hudson, Deavers, and Bradley (1972) on desert ground squirrels of the genus *Citellus* in the deserts of the southwestern United States.

Ground squirrels are day-active; some hibernate (*Citellus armatus, C. spilosoma,* and *C. townsendi*), and some do not (*C. leucurus* and *C. tereticaudus*). Hibernation and estivation are physiological states that allow avoidance of critically high temperatures in summer and low temperatures in winter. However, as Hudson, Deavers, and Bradley (1972) point out, it requires only a slight physiological shift to allow a hibernator to adapt to a tolerance of desert conditions—that is, to tolerate prolonged exposure to high ambient temperatures. Desert ground squirrels, such as *Citellus leucurus, C. spilosoma, C. townsendi,* and *C. tereticaudus,* can survive a two-hour exposure to temperatures of 40°C and above, while montane species cannot. But there are more subtle differences, particularly within desert species. *C. spilosoma* from the high plateau desert of New Mexico has an upper lethal temperature of 40°C, whereas *C. tereticaudus*, which lives in habitats characterized by extreme heat and aridity, has an upper lethal temperature (measured over a two-hour period) of 46°C.

These observations suggest some degree of tolerance of body temperature fluctuation, variously expressed depending on the amount of environmental stress to which a species is exposed. There are parallels here with desert marsupials. Dasyurids, such as *Dasycercus cristicauda, Sminthopsis froggatti*, and *S. crassicaudata*, and the peramelid *Macrotis lagotis* (the bandicoot) show a diurnal cycle of body temperature fluctuation in Australian deserts (Tyndale-Biscoe 1973). *D. cristicauda* is a nocturnal, surface-active carnivore that lives in a subterranean burrow during the day. Here it becomes torpid, and its body temperature drops as much as 3°C to 4°C. *Sminthopsis froggatti*, which inhabits the same sandy, arid areas as *D. cristicauda*, has much the same behavior pattern. In contrast the fat-tailed marsupial mouse, *S. crassicaudata*, has a body temperature that fluctuates less than 1°C over the diurnal cycle, and this species rarely becomes torpid. Indeed, it cannot afford to do so since it is smaller than *S. froggatti* and takes smaller items of prey. It has to forage more extensively for its food, although it has a fat store in its tail that may enable it to survive short periods of food stress. Further, *S. crassicaudata* lives in more shaded sites than the other two species, where thermal stress is likely to be less severe.

The story does not end here, however, for there is evidence that some species of desert rodents have adopted an alternative strategy to that which involves the tolerance of a high and/or fluctuating body temperature. This alternative is, in effect, thermoregulation. McNab and Morrison (1963) identified thermoregulation in various subspecies of *Peromyscus eremicus* and *P. crinitus* inhabiting desert environments in the United States. These rodents are mainly crepuscular but can extend their activity on the soil surface to periods in the afternoon when ambient temperatures may exceed 35°C. *P. crinitus* exhibits thermoregulation at a temperature of 38°C and is better, in this respect, than its mesic relatives. This ability to regulate body temperature at high ambient temperatures is coupled with an inability to regulate at low ambient temperatures. This may be a decisive factor limiting the distribution of these rodents to hot deserts.

Metabolic Compensation. The assumption of a torpid state and the attendant depression of the body temperature during the hottest part of the day by *Dasycercus cristicauda* and *Sminthopsis froggatti* are ways of conserving energy since the metabolic rate will be lowered. Another way in which this energy conservation can be achieved is by becoming heat adapted; isopods, the wolf spider *Lycosa carolinensis*, and the millipede *Orthoporus ornatus* illustrate this.

Oxygen consumption in the Sonoran wolf spider *Lycosa carolinensis* is stabilized within 72 hours of exposure to a temperature increase from 22°C to 39°C (Moeur and Eriksen 1972; see also Crawford 1979a). This rapid temperature acclimation is complemented by metabolic compensation in the temperature range of 39°C to 45°C. Semixeric (and presumably xeric) isopods show temperature acclimation in regard to oxygen consumption, also. Both Cloudsley-Thompson (1969), working with *Metaponorthus pruinosus* and *Periscyphis jannonei*, and Edney (1964b), investigating *Armadillidium vulgare* and *Porcellio laevis*, showed that oxygen consumption declined markedly at temperatures in the upper range of tolerance of these species. This indicates that rates of energy expenditure decline at these temperatures, but it was left to Wooten and Crawford (1974) to suggest how this might be achieved as a result of their studies on *Orthoporus ornatus*. These authors demonstrated that *O. ornatus* shows metabolic compensation at temperatures between 25°C and 35°C. Within this range respiratory Q_{10} values (RQs) were significantly lower than values obtained at temperatures in the 15°C to 25°C range. This suggested that energy expenditure is relatively reduced in hot weather. This would allow *O. ornatus* to remain relatively active and feeding in microsites where, in the absence of metabolic compensation, it would undergo severe heat stress due to increased heat loading (from metabolic reactions) and body water loss (from increased respiration). The mechanism involved in this metabolic shift would seem to be a utilization of carbohydrate, in preference to fat, substrates during the summer season, with the reverse occurring during the winter. The determination of RQ values for summer and winter individuals indicates as much. During the summer period, RQ values of 0.90 were recorded, and this suggests that carbohydrate substrates were being metabolized. At this time *O. ornatus* was feeding and, as a consequence, receiving an adequate supply of carbohydrates from *Ephedra* bark and creosote bush litter in the Chihuahuan desert. In winter lower RQ values of 0.77 suggested that fatty storage products were being metabolized to provide sources of energy for quiescent individuals. *Orthoporus ornatus* is a discontinuous feeder, when this activity is considered on an annual basis, and it is of interest to record that Wallwork (1975), using an entirely different approach based on bomb calorimetry, identified precisely this same type of feeding strategy in discontinuous feeders among the soil fauna of a mesic beech wood.

8
Species Diversity and Resource Allocation

It is almost axiomatic that biological activity in hot deserts is very much lower than it is in mesic environments. The extremes of the physical environment are the factors that limit biological activity and against which survival strategies have to be developed: high and low temperatures, drought and unpredictable floods, and unstable soils. The line between survival and extinction is so finely drawn in these demanding environments that competition is a luxury few species can afford. Instead, various animal and plant populations associate with each other and may even cooperate to ensure their collective survival. In addition mechanisms exist for allocating resources to minimize direct competition and, thereby, maintain a level of species diversity, which for some groups at least is surprisingly high.

BIOTIC ASSOCIATIONS

As we have already noted, the millipede *Orthoporus ornatus*, honey-pot ants (*Myrmecocystus* spp.), certain Collembola, and snails (Fig. 8.1) are at least partly dependent on the aerial parts of desert shrubs for refugia and/or food. Similarly the iguana *Dipsososaurus dorsalis* relies on the shade of the creosote bush to protect it from the heat of the day; when surface temperatures become extreme, this lizard climbs among the foliage of this desert shrub. Beneath the soil surface kangaroo rats (*Dipodomys* spp.) and

Fig. 8.1. The snail *Sphincterochila* among the foliage of *Artemisia herba-alba* in the Negev desert. (Photo by C. S. Crawford.)

pocket mice (*Perognathus* spp.; Fig. 8.2) frequently locate their nest chambers in the leaf catchment zone beneath the rooting systems of shrubs such as the creosote bush. Already, we have had occasion to refer to the creosote bush (Chapter 5) as a major source of organic input to hot desert soil systems. Now we see it in another important role—as a structural feature in the environment of various desert animals, one that is probably vital to their survival. These associations between animals (such as millipedes, ants, springtails, snails, lizards, and rodents) and desert shrubs are examples of one-way traffic; they benefit the animals but not the plant. *Orthoporus ornatus* also obtains a one-way benefit by virtue of its ability to

Fig. 8.2. The pocket mouse *Perognathus longimembris*, San Bernardino County, California. (Photo by R. E. MacMillen.)

retreat into the subterranean nests of the harvester ant *Novomessor* (see Fig. 2.6); desert thysanurans use the nests of ants and termites in a similar symbiotic fashion.

Other associations in hot deserts can be cited in which both plant and animal benefit; a case in point is the well-known mutualism that occurs between the moth *Pronuba* and the *Yucca*, but this is an association that does not fall within the confines of the soil system. Perhaps the closest subterranean parallel is provided by granivorous ants and rodents that harvest seeds above ground and then deposit these in subterranean caches. Many of these seeds undoubtedly provide food reserves for the animals, but some may be enabled to germinate by being "planted" in a moist environment below ground; there seems to be no hard evidence for or against this possibility, however. Similarly, the deposition of feces in underground nests potentially may enrich the soil in the vicinity of rooting systems of the shrubs that provide sheltered locations for these nests (see above). These associations—some cooperative,

192 Desert Soil Fauna

others unconsciously altruistic—may be vital to the survival of a species in hot deserts and contribute to the general biotic diversity.

However, it is not with biotic diversity on a global scale that this chapter is concerned. Adaptive radiation has occurred to a marked degree within certain taxonomic groups or feeding guilds, such as the ants and the rodents, and high levels of species diversity are the consequence. Before we examine this diversity and the mechanisms that sustain it, it is necessary to define the term *species* in the context of the desert soil fauna.

THE ECOLOGICAL SPECIES

Many of the groups of desert soil animals discussed in the previous chapters present few taxonomic problems as far as the identification of morphospecies is concerned. Ants, tenebrionid beetles, millipedes, centipedes, scorpions, and other large arachnids, woodlice, and mollusks, for example, can readily be identified to species by morphological criteria. Not so, however, with certain groups of the subterranean microfauna, such as the mites and nematodes. Prostigmatid mites in particular, which are abundant and widespread in desert soils, provide serious taxonomic problems since many families, genera, and species are very incompletely known. In addition it is often difficult to match juvenile stages with adults in soil faunal extracts, and this is also true to some extent for cryptostigmatid and mesostigmatid mites. To postpone the ecological study of these groups until their taxonomy has been fully worked out would be to postpone such study indefinitely. To circumvent this, we can recognize that the juveniles and adults of any given morphospecies of soil mite may occupy different niches and can therefore be treated as different ecological species. Similarly, adults of different morphospecies are considered ecologically distinct, even though the species themselves may not have been described and named.

SPECIES DIVERSITY

It is easy enough to generalize, and thereby underestimate, species diversity in hot desert soils. Our preoccupation with the rigors of the desert environment and their impact on the soil fauna should not obscure the fact that a number of faunal groups have exploited the hot desert soil system in a relatively successful manner and, as a result, achieved high species diversity. Ants and rodents have already been mentioned in this context; tenebrionid beetles also spring immediately to mind. A less well-known example relates to the more truly subterranean fauna of mites, Collembola, psocopterans, japygids, insect larvae, pauropods, pseudoscorpions, and spiders occurring in a range of soil habitats in the Chihuanhuan desert (Wallwork, in preparation). Here, diversities approaching those of mesic temperate forest soils have been recorded (Table 8.1). It must be pointed out, however, that the highest diversity values in these sites were found in sheltered organic soils that are essentially mesic in character.

The intriguing question is how these levels of faunal diversity can be maintained in the face of limited and discontinuous resources. In theory mechanisms must exist for the differential allocation of the available resources. But what is the nature of these mechanisms, and how do they operate? There is no single, simple answer to these questions. The outcome depends very largely on the animals in question; on sympatry and allopatry, which is best examined in the context of habitat separation; and on niche breadth and species packing, which are perhaps more a function of feeding tactics. As we will see, these parameters are often difficult to separate. Perhaps we can make a start by exploring the possibility that patterns of resource allocation among the subterranean fauna may be different from those above ground.

RESOURCE ALLOCATION: THE SUBTERRANEAN FAUNA

Although the subterranean fauna is primarily invertebrate in character, it does include a vertebrate component that is mainly represented in hot deserts by the fossorial mammals. This component is considered first.

TABLE 8.1.
Diversity Parameters for Total Ecological Species of Soil Microarthropods from Four Chihuahuan Desert Sites

Site	Organic Content (%)	Species Diversity (H')	Equitability (e)
Juniper	40.4	4.73	0.44
Saltbush	28.4	4.04	0.48
Creosote bush	8.6	3.45	0.25
Black grama grass	4.0	1.10	0.15

Source: Wallwork, unpubl.

Mammals

Eisenberg (1975) classified the subterranean rodents as herbivores, and these are well exemplified in North American deserts by pocket gophers of the family Geomyidae. The other main group of fossorial mammals, the Insectivora, has not achieved anything like the success of the rodents in these deserts (Nevo 1979). However, insectivorous moles do occur in the cool, arid parts of Asia and enjoy a widespread distribution here. High intra- and interspecific competition occurs among and between rodents, and territorial behavior is developed to minimize encounters between individuals and species. Nevo (1979) makes the observation that limited resources in the underground environment result in low species diversity in a given area and the selection of specific microhabitats by sympatric species, although he does not provide examples. He further points out that keen intraspecific competition leads to populations that are relatively small, subdivided, semiisolated, and territorially structured. Substantial overlaps in distribution of subterranean mammals only occur between herbivores (rodents) and insectivores (moles), which are obviously not competing for the same food resources. These two groups can coexist by virtue of the differential allocation of resources, in the broadest sense.

Microfauna

Soil communities and their constituent food webs are based for the most part on detritivore food chains. The term detritivore in this context usually includes species that feed not only on the debris from higher plant material but also on algae, fungi, and bacteria that may be ingested discriminately or indiscriminately along with organic detritus. To this trophic category belong many cryptostigmatid and prostigmatid mites, various Collembola, Psocoptera and dipteran larvae, and nematodes. The diversity of this microfauna in desert soils has already been remarked upon earlier in this chapter, and it is pertinent to scrutinize this more closely now. Table 8.2, for example, compares the numbers of species of the three main groups of mites occurring in various Chihuahuan desert soils. The immediate impression, from a perusal of these data, is of a remarkable diversity within the detritivore level. The two mite groups that contribute to this level, the Cryptostigmata and Prostigmata, collectively account for 80 percent of the species and more than 85 percent of the individuals in the mite fauna of the juniper

TABLE 8.2.
Numbers of Individuals and Species and Species Diversity of the Three Main Groups of Soil Mites in Four Chihuahuan Desert Sites

	Cryptostigmata			Prostigmata			Mesostigmata		
	N	S	H'	N	S	H'	N	S	H'
Juniper	1,204	34	3.61	222	22	3.14	240	14	2.74
Saltbush	108	10	2.74	653	8	1.80	250	12	2.76
Creosote bush	321	12	1.16	103	22	3.61	42	6	1.77
Black grama grass	449	8	0.46	7	5	2.23	22	1	–

N = Number of individuals
S = Number of species
H' = Species diversity
Note: Values for N and S are totals from five sampling units.
Source: Wallwork, unpubl.

site, for example. The diversity illustrated by these data is by no means peculiar to desert soils; ecologists working in mesic temperate soils have been attempting an explanation of this phenomenon for some time (Anderson and Healey 1972; Anderson 1975; Wallwork 1976).

However, the situation is accentuated in deserts since the organic base for the detritivore food chain is discontinuously distributed. Plant litter accumulates immediately beneath desert shrubs, and these accumulations are separated from each other by bare and inhospitable expanses of mineral soil (Fig. 8.3). These accumulations represent biological islands of varying sizes, and faunal exchanges between them are minimal. As a consequence of this isolation there is a real likelihood of potentially competitive encounters not only at the detritivore level but also between carnivores as well. The mechanisms that ensure that such competition does not operate to reduce species diversity to a minimum are not well understood, although there is some circumstantial evidence that suggests that the substitution of ecological vicariants may occur from place to place. As we have already noted in Chapter 4, there is a reciprocal relationship between the relative abundances of cryptostigmatid and prostigmatid mites in Chihuahuan desert soils. Evidently detritivorous prostigmatids, such as the tarsonemids, flourish in mineral soils where detritus-feeding cryptostigmatids are poorly represented; the reverse is the case in organic soils. This may be the outcome of interference competition, but the more likely explanation is a passive one. The Prostigmata that live in

Fig. 8.3. Discontinuous pattern of surface vegetation in the Joshua Tree National Monument, Mojave desert.

soil and feed on detritus and/or microorganisms are small in body size and may be able to move through the pore spaces between mineral soil particles with greater facility than the larger-sized Cryptostigmata. This would give them an advantage in colonizing mineral soils. On the other hand, the fungivorous habit of many cryptostigmatids will direct them into selecting organic soils where fungal populations are well developed and soil pore spaces are larger. Thus, different habitat preferences may account for the reciprocity between these two groups.

The Prostigmata also includes a number of families the members of which are predators. Those that occur in hot desert soils, such as the Bdellidae, may be brought into competitive encounters with predatory mesostigmatid mites, which also occur here.

However, as noted in Chapter 4, this competition may be minimized by the predilection of the Prostigmata for arid sites; the Mesostigmata for sites with a rather higher moisture content. This idea receives further support from the data presented in Table 8.2, which show that although the species diversity of Mesostigmata is not very great in the four sites, it is highest in the two organic soils (juniper and saltbush). These soils will have a greater water-holding potential than the mineral soils under creosote bush and black grama grass.

In a very general sense, then, it would seem that the nature of the substratum (that is, whether it is organic or mineral) and the mean annual rainfall determine the relative contributions of the Prostigmata and the Cryptostigmata to the detritivorous fauna; they also determine the Prostigmata/Mesostigmata ratios among the predatory element. Different habitat preferences reduce the possibility of these ecological groups entering potentially destructive competitive situations. Admittedly, this is a very superficial interpretation, which glosses over the specific mechanisms that promote species diversity among the mite component of the soil fauna in any particular desert site. Truth to tell, we are very largely ignorant of the way in which these mechanisms function not only in desert soils but also in the much-studied mesic temperate soils. There are several questions that need to be addressed in this context. Does competition exist between, for example, detritivorous Prostigmata and Cryptostigmata, or between different groups of predatory mites in hot desert soils? Certainly, there are factors that minimize this competition, but it is unlikely that they can eliminate it completely. What other back-up systems are available? To answer this question, it is necessary to focus attention on the effective microhabitat of soil organisms: the constellation of factors in their immediate environment to which they will immediately respond. If we do this, we find that potential competitors may be separated not only by different habitat preferences on a geographical scale but also by their response to the diversity of microhabitats within a particular soil profile. This diversity has structural, microclimatic, and biochemical components (Wallwork 1976). Exactly how the various elements of the soil fauna respond to this diversity remains another unanswered question. By contrast we know much more about the ways in which habitat selection and resource allocation influence species diversity in that part of the soil fauna that is active on the ground surface, and it is to this that we now turn.

RESOURCE ALLOCATION: THE SURFACE-ACTIVE FAUNA

Three major trophic groups can be identified in the surface-active fauna of hot deserts, namely, the carnivores, detritivores, and granivores. Each of these is now considered in turn.

Carnivores

It is not uncommon in hot deserts to find two or more groups of large-sized predatory arachnids living sympatrically. The uropygid *Mastigoproctus giganteus* has a distribution in deserts of the southwestern United States that overlaps with those of various scorpions. Solifugids and scorpions inhabit the same regions in African deserts. Naturally, the question is raised as to how these predatory, and potentially competitive, groups can coexist. One possible answer lies in the inherent aggressive qualities of these arachnids. Cloudsley-Thompson (1961c) demonstrated that the camel-spider *Galeodes arabs* and the scorpion *Leiurus quinquestriatus* will fight to the death when they encounter each other. The camel-spider usually emerges as the victor from these encounters, particularly if it is a female. Such aggression may serve to keep different species populations spatially separated under natural conditions and thereby enable them to exploit different prey pools.

More subtle mechanisms of separation evidently are employed by the scorpions *Centruroides sculpturatus* and *Diplocentrus spitzeri*. These two species, which occur sympatrically in southwestern New Mexico, live under rocks. *D. spitzeri* excavates shallow burrows in the soil, whereas *C. sculpturatus* tends to hang on the undersides of rocks or seek shelter in rock or bark crevices; hence, the common name of bark scorpion. *D. spitzeri* forages for food in the natural darkness of its rock shelter. Here it feeds, in particular, on tenebrionid beetles. In contrast *C. sculpturatus* has a natural endogenous rhythm of activity that stimulates it to forage on the surface at night. Thus, resource allocation is brought about by different temporal and spatial foraging strategies in the case of these two predators (Crawford and Krehoff 1975). *D. spitzeri* may be regarded as a continuous, subterranean feeder, whereas *C. sculpturatus* is a discontinuous, surface feeder.

The way in which different patterns of feeding activity can operate to reduce competition between two predator species is also well illustrated by horned lizards (*Phrynosoma* spp.) in the Chihuahuan desert. Here, two species, *P. modestum* and *P. cornutum*, are sympatric and both feed on ants. As we have seen in Chapter 3, *P. modestum* feeds for the most part on honey-pot ants (*Myrmecocystus* spp.), whereas *P. cornutum* preys exclusively on members of the harvester genus *Pogonomyrmex*. This differential allocation of the "ant" resource is brought about by different feeding patterns of the two lizards. *P. cornutum* is a wandering forager (Whitford and Bryant 1979), traveling across open areas in search of its prey. It takes large-sized ants (greater than 5 mm), and feeding activity is at its peak during the late morning hours. At this time the only suitable prey available are *Pogonomyrmex* spp. In contrast *P. modestum* is a sedentary lizard, feeding under the shelter of vegetation. Here it is well situated to prey on *Myrmecocystus* spp., which move between the shrub canopy and soil at times of the day when *P. modestum* is active (Shaffer and Whitford, 1981).

The ant fauna of hot deserts also includes a carnivorous or arthropod-foraging guild, and the ways in which different species belonging to this guild can coexist have been studied by Chew and De Vita (1980). These authors identified the species *Iridomyrmex pruinosum*, *Conomyrma insana*, and *Myrmecocystus depilis* as arthropod foragers that coexist in the desert of southern Arizona. These three species select different sizes of prey, and this selection is directly related to their body length and the functional gape of their mandibles (Table 8.3). This same phenomenon is evidently also a feature of resource allocation in coexisting granivorous ant species (see below).

Detritivores

Three groups of surface-active detritivores can be singled out for special mention here: snails, tenebrionid beetles, and termites.

TABLE 8.3.
Relationship between Body Size, Mandible Gape, and Forage Size in an Arthropod-Foraging Guild of Desert Ants

Species	Body Length (mm)	Mandible Gape (mm)	Forage Size (mm)
Iridomyrmex pruinosum	2.13	0.20	1.76
Conomyrma insana	2.93	0.28	1.80
Myrmecocystus depilis	4.44	0.42	14.10

Source: Adapted from Chew and De Vita 1980.

Snails. The five species of desert snails belonging to the genus *Sphincterochila* (see Chapter 2) have distributions in the arid and semiarid parts of the Mediterranean region that, for the most part, do not overlap. This spatial separation is brought about mainly by different moisture and substrate preferences (Table 8.4) and, only secondarily in regions of low rainfall, by slope (aspect). *Sphincterochila cariosa* is limited to north coastal and inland areas of Israel, which receive an adequate amount of rainfall and provide calcareous substrata; *S. fimbriata, S. zonata,* and *S. prophetarum* occur inland, the first in semiarid conditions, the other two in rather drier areas. *S. aharonii* is restricted to coastal regions (Bar 1975). The aspect factor is only important in the arid mountainous region of central Israel where the distributions of *S. zonata* and *S. prophetarum* overlap. Here, separation is achieved by the preference of *S. zonata* for loessial plains, which constitute about 13 percent of the total area of the Negev, while *S. prophetarum* lives on rocky slopes, which occupy about 42 percent of this desert (Yossi Steinberger 1980: personal communication).

Tenebrionids. This is a remarkably diverse family in hot deserts. In the dune desert of the Namib, for example, Edney (1971) recorded seven species belonging to the genera *Onymacris, Gyrosis, Calosis,* and *Lepidochora.* The diversity of the tenebrionid fauna in this desert can probably be explained in spatial and temporal terms. The Namib is a dune desert dissected by riverbeds and gravel plains. The dunes themselves offer several subhabitats—some with vegetation (dune bases and interdune valleys); others

TABLE 8.4.
Parameters Affecting the Distribution of Snails of the Genus *Sphincterochila* in the Deserts of Israel and Sinai

Species	Isohyetal Boundaries (mm)	Substrate
S. cariosa	1,000–500	Limestone and dolomite with terra rosa soil, chalk crust, and calcareous sandstone
S. fimbriata	500–200	Rendzina or brown earth on chalk
S. aharonii		Exclusively on calcareous sandstone
S. zonata	200–100	Loess and flint on chalk
S. prophetarum	100–50	Loess and desert lithosols on rocky slopes

Source: Compiled by the author.

that lack vegetation (dune slopes, crests, and slip faces). This desert offers a spatial heterogeneity that can be exploited by tenebrionids. Different species populations are physically separated by selection of different subhabitats. This separation is reinforced by different patterns of activity, both seasonal and diurnal (Table 8.5). Members of the genus *Lepidochora* are nocturnal, while the remainder, with the possible exception of the crepuscular *Onymacris laeviceps*, are active during the daytime. These day-active forms are also summer beetles and as such are exposed to the driest and hottest conditions. That they can survive these conditions is largely due to a very impermeable cuticle and spiracular closing mechanisms, which reduce water loss from the body to a minimum. The crepuscular *O. laeviceps* loses water at a greater rate than the diurnal, summer beetles *Onymacris plana* and *Calosis amabilis* but at a lower rate than the nocturnal, summer- and winter-active *Lepidochora porti* and *L. argentogrisea* (Edney 1971). It seems quite clear, therefore, that tenebrionids respond to environmental water stress in different fashions. These differences may provide the foundation for a selective process that minimizes interspecific encounters and subsequent competition for the detritus food source.

In short the tenebrionids that have the greatest ability to conserve water—that is, those that can withstand the greatest environmental stress—are those that are active during the most

TABLE 8.5.
Activity Patterns and Habitat Preferences of Various Species of Tenebrionid Beetles in the Namib Desert

Species	Activity				Habitat
	Winter	Summer	Day	Night	
Onymacris plana	+	+	+	−	Dune and slip face bases
O. laeviceps	−	+	Twilight		Dunes and dune bases
O. rugatipennis	+	+	+	−	Riverbed gravel
Gyrosis moralesi	−	+	+	−	Interdune valleys
Calosis amabilis	−	+	+	−	Gravel plains
Lepidochora porti	+	+	−	+	Dune crests and slip faces
L. argentogrisea	+	+	−	+	Dunes and dune bases

+ = Active
− = Inactive
Source: Data from Edney 1971.

demanding times: they are day-active and summer-active. The species with a lesser ability to conserve body water restrict their activity to the hours of darkness when temperatures are lower and moisture is available in greater amounts. There is, then, a temporal separation of related species that are potential competitors. But this is refined even further. A species such as *Onymacris laeviceps*, which loses water at a rate that is intermediate between that of the day-active and nocturnal groups, fits neatly into a crepuscular pattern of activity, which again, minimizes competition with other tenebrionids.

Interspecific encounters are further minimized by spatial separation of populations (Table 8.5). Two of the most conservative tenebrionids in the Namib, as far as their ability to curtail water loss is concerned, are *Onymacris plana* and *Calosis amabilis*. Both of these species are day-active and summer beetles. Yet competition between them is reduced by *O. plana* selecting dune habitats and *C. amabilis* preferring gravel plains. The dune habitat would appear to present a more inhospitable environment than the gravel plains, and the success of *O. plana* here may be due to its relatively large body size and the consequent low surface area/volume ratio, compared with *C. amabilis*. Two other day-active, summer tenebrionids in the Namib are *Onymacris rugatipennis* and *Gyrosis moralesi*, which are spatially separated from each other, and from *O. plana* and *C. amabilis*, by their different habitat preferences—*O. rugatipennis* for riverbed gravels, and *G. moralesi* for interdune valleys.

These patterns of tenebrionid distribution in hot deserts illustrate what is perhaps a fundamental concept in ecology, namely, the existence of refuges. Implicit in this concept is the assumption that species distributions in nature overlap to a greater or lesser extent but that there is always a part of the range, or more broadly, the niche, of each species in which it is better adapted than are other species. The niche has both spatial and temporal dimensions; it is not only, or merely, a physical space in the environment. It can include a characteristic pattern of feeding behavior, the timing of peak periods of activity, as well as differential responses to the structural, microclimatic, and biochemical components of habitat diversity. In theory the niche has an infinite number of dimensions, but in hot deserts gradients of physical factors, and species' response to these gradients, achieve special significance. The tenebrionids would appear to illustrate this point very well.

This preoccupation with the physical effects of the hot desert environment, widespread though it may be, must not allow us to overlook the fact that some of the most successful colonizers of arid lands have achieved such a measure of independence from their abiotic environment that they are not constrained by it. This independence has been gained either by the creation of an insulated microcosm—in more real terms, a subterranean nest or surface mound—or by a combination of behavioral and physiological traits that minimize or eliminate environmental stress. The former strategy is employed by termites and ants; desert rodents, on the other hand, use both of these strategies.

Termites. It is not stretching the concept of the detritivore niche to unacceptable limits to mention the termites at this point. The detritus on which they feed, notably dead wood and dung, comes in larger packages than that exploited by the tenebrionids, although some termites also feed on particles of herbaceous leaf litter. These social insects, with the exception of the Macrotermitidae, have the ability to digest cellulose in appreciable quantities, and this is militated through the activities of a gut fauna of symbiotic protozoan flagellates.

The use of cellulose as a food base opens up many feeding opportunities for termites, and this factor may be an important determinant of their success in and around the margins of hot deserts. The cellulose in wood, dung, and leaf litter is differentially exploited by different feeding guilds of termites (Watson, Lendon, and Low 1973), and this provides a mechanism for resource allocation that minimizes interspecific competition. This is reinforced by different spatial preferences shown by sympatric populations that separate species constructing surface mounds from those that live in subterranean nests. For example, the mounds of the Australian *Tumulitermes tumuli* are often situated in the shade of low shrubs (Lee and Wood 1971), whereas the nests of subterranean species occur more frequently in open areas.

Granivores

It will be evident from what has been said in Chapter 5 that desert systems can be conveniently described by the pulse-reserve paradigm, and at the primary producer level the reserve element in this module is very largely represented by seeds. It is not surprising, therefore, that ecologists interested in the functioning of desert systems have paid considerable attention to seed-feeding, or granivorous, animals. These seed predators, as they are sometimes called, belong to three main groups: the ants, rodents, and birds. The last group falls outside the scope of this book since they are not soil animals, but the other two groups are, and they are now dealt with in turn.

The species diversity of granivorous, or harvester, ant faunas in the deserts of the southwestern United States has been investigated by Davidson (1977b). She established 10 study sites on a longitudinal gradient of increasing rainfall from Baker, California (Mojave desert), through Gila Bend, Ajo, and Casa Grande, Arizona (Sonoran desert), to Rodeo and Deming, New Mexico (Chihuahuan desert). The relationship between mean annual rainfall and species diversity along this gradient is positive (Table 8.6). The total species pool of ants in these deserts was at least 15, contributed by five myrmicine genera, and local communities along the gradient included between 2 and 7 seven common species. Similarly, Chew and De Vita (1980) recorded six species of granivorous ant (*Pheidole rugulosa*, *Ph. xerophila tucsonica*, *Ph. militicida*, *Pogonomyrmex imberbiculus*, *P. desertorum*, and *P. rugosus*) in an assemblage in southern Arizona. Naturally, the question arises: How do two, three, four, up to seven granivorous ant species exist sympatrically? There are several possible answers to this question, some simple, some more complicated. For example, interference competition may bring about spatial separation of nesting sites and foraging territories of congeners (Brown, Reichman, and Davidson 1979). Food preferences may also be important. Davidson (1977b) examined the preference of different ant species for seeds of varying sizes, and she found that seed-size preferences were highly correlated with the body sizes of the worker ants: the foraging caste. This phenomenon was also observed in the case of members of the genus *Messor* in the Egyptian coastal desert (Ghabbour,

TABLE 8.6.
The Relationship between Mean Annual Precipitation and the Species Diversity of Ant Faunas in the Deserts of the Southwestern United States

Locality	Mean Annual Rainfall (mm)	Species Diversity (H')
Baker, California	76	0.59
Mojave, California	121	0.14
Gila Bend, Arizona	142	1.08
Ajo, Arizona	216	1.43
Casa Grande, Arizona	215	1.12
Rodeo, New Mexico	276	1.67
Deming, New Mexico	224	1.54

Source: Data from Davidson 1977b.

Mikhail, and Rizk 1977) and in the study carried out by Chew and De Vita mentioned above. The latter authors also showed that mandible gape was correlated with seed-size preference (Table 8.7). All this provides a mechanism whereby species of different body size can coexist because they utilize different food sources. Furthermore, Davidson (1977b) found that species of similar body size could also coexist provided that they have different foraging strategies.

This concept of foraging strategies needs to be scrutinized more closely in relation to granivorous ants in hot deserts. Whitford and his co-workers in New Mexico have reported a considerable overlap in seed selection between large-sized *Pogonomyrmex* spp. and small-sized *Pheidole* spp.; the ants of both of these genera concentrate on the seeds of buckwheats, annual grama grasses, and fluff grass in the Chihuahuan desert (Whitford et al. 1981). This contradicts the findings of Davidson and others outlined above and places greater emphasis on temporal, as opposed to spatial, segregation to minimize interspecific competition. *Pogonomyrmex* spp. are primarily diurnal foragers, as we have seen, whereas *Pheidole* spp. are mainly crepuscular (Whitford and Ettershank 1975; Whitford 1978b; Whitford et al. 1981). Further, mem-

TABLE 8.7.
Relationship between Body Size, Mandible Gape, and Forage Size in a Seed-Foraging Guild of Desert Ants

Species	Body Length (mm)	Mandible Gape (mm)	Forage Size (mm)
Pheidole rugulosa	1.64	0.19	0.52
Ph. xerophila tucsonica	1.96	0.25	1.29
Pheidole militicida	2.69	0.33	2.19
Pogonomyrmex imberbiculus	4.13	0.42	4.07
Pogonomyrmex desertorum	5.34	0.63	10.10
Pogonomyrmex rugosus	7.22	0.87	10.90

Source: Adapted from Chew and De Vita 1980.

bers of these two genera show a seasonal separation of foraging activity, with *Pogonomyrmex* spp. showing peak activity in early and midsummer in the Chihuahuan desert. In contrast *Pheidole rugulosa*, *Ph. xerophila*, and *Ph. militicida* were highly active in October and early November, foraging on such late-seeding plants as *Erioneuron pulchellum* at a time when *Pogonomyrmex* spp. were largely inactive (Whitford et al. 1981). A further dimension is added to this picture if patterns of resource density are taken into consideration.

Ants forage for seeds either individually or in groups. Group foraging is characteristic of species living in mesic environments, and it is related to high density resources (Davidson 1977a). In the deserts of the southwestern United States, such a strategy is employed by *Pheidole militicida*, *P. xerophila*, *P. gilvescens*, *P. sitarches*, *Pogonomyrmex barbatus*, and *Solenopsis xyloni*. These species forage during the most favorable periods of the year; they store seeds in their nests and pass the winter and summer in hibernation and estivation. Some desert ants, such as *Novomessor cockerelli*, *Pheidole desertorum*, and many *Pogonomyrmex* spp., have adopted the strategy of foraging individually, and this appears to be a response to low density resources. *Pogonomyrmex rugosus* is unusual in having a mixed foraging strategy (Whitford 1976a), and group activity is highest during times of peak seed abundance.

Seed production has a large, unpredictable element in its timing and magnitude in hot deserts. This follows naturally from the fact that it is a sequel to flowering that is triggered by rainfall events that are stochastic. Such a system will provide high density resources for seed feeders at certain times and low density resources at others. Coupled with this is the fact that group foragers store food against times of potential food stress. These are the times when individual foragers may be active. Herein lies a possible mechanism for the temporal separation of potential competitors; it may be incomplete when activity patterns overlap, but it may also be strengthened by spatial separation. Once again, the rainfall factor is implicated.

Seeds of desert plants are relatively small and do not as a rule remain in situ on the soil surface after being liberated. They may be dispersed widely by the high winds that are often characteristic of hot deserts. More important, they are gathered up in the rainwater streams that course along wadis and arroyos, and they are deposited in varying concentrations on the strand lines of these washes and around playa lakes. Here, matted together with fragments of stems and leaves, they achieve a measure of permanence.

This transportation process results in the occurrence of a wide range of seed distribution, which may be unrelated to the actual amount of seed fall. There is, then, a spatial separation of foci of high and low density resources over a given area at any one time. This allows for species with different foraging strategies to operate at the same time without coming into competition.

It seems likely, therefore, that the diversity of the ant fauna of hot deserts cannot be explained solely in terms of a response to one variable. Differential preferences for seed size operate alongside different foraging strategies that have both temporal and spatial dimensions. In a more general sense the costs and profits of foraging species are defined by physical and biotic parameters, such as food availability, capacity for food storage, hunger, surface temperatures, and saturation deficits, which produce an activity pattern governed by an integration of responses to these parameters (Whitford and Ettershank 1975). In simpler terms foraging activities expend energy, and this expenditure must be balanced by the rewards. Each granivorous species has its own balance sheet, its own energy budget, which is controlled by the ability to acquire food, to store this food, and to utilize it when necessary. These functions are directly related to the nature and intensity of foraging activity, which is controlled by hunger, the abundance of food, and the microclimate at the soil surface (see Brown, Reichman, and Davidson 1979 and Whitford et al. 1981 for further discussion and an introduction to the literature relating to this topic).

Ants and rodents would seem to make strange bed-fellows, but the patterns of species diversity and community organization among the seed-feeding representatives of both of these groups show great similarities (Davidson 1977b; Brown, Reichman, and Davidson 1979). These similarities are of intrinsic interest to the ecologist since desert ants and rodents are often sympatric. Both of these unrelated groups are the main granivores in hot deserts, and they operate in a pulse environment where their food supply (seeds) is unpredictable in its frequency and magnitude. Both include species that have adopted similar strategies of foraging, food storage, hibernation, and estivation to minimize environmental stresses. That they exist in a stable, competitive relationship is suggested by the finding that ant densities increase by 71 percent in desert plots when rodents are excluded, and rodent densities increase by 20 percent when ants are excluded (Brown, Davidson, and Reichman 1979). The mechanisms that allow this competitive relationship to exist will be considered later. For the moment we are concerned only with the rodents.

The species diversity of the rodents inhabiting the deserts of the southwestern United States is at least as great as that of any of the other deserts of the world, and without question this is the area that has been most thoroughly studied. In these deserts five or six granivorous rodent species can be encountered in the same area, and their ability to coexist requires an explanation. Similarly, the deserts of Africa and Asia harbor a diverse granivorous rodent fauna that shows many convergences, notably in patterns of body size and foraging behavior, with that of North American deserts (Brown, Reichman, and Davidson 1979). In these African and Asian deserts the problems of coexistence are the same as in the North American deserts. In attempting to solve these problems granivorous desert rodents have adopted two different, but not unrelated, strategies: habitat selection and resource allocation.

Habitat Selection. Deserts provide a spectrum of habitats with horizontal and vertical dimensions. Desert rodents exploit both of these components of habitat diversity, and we can now examine the ways in which they do this.

Some of the rodents inhabiting the Rajasthan sector of the Thar desert of India show marked habitat specificity, which expresses itself in the horizontal separation of potential competitors. *Gerbillus gleadowi* and the related *G. nanus*, for example, are spatially separated, the former selecting sand dunes in which it constructs shallow burrows, the latter preferring sandy plains. Further spatial separation is shown by the rock-inhabiting *Rattus cutchicus* and *Mus cervicolor*. The most common rodent in this region is the gerbil *Meriones hurrianae*, a species that is less specific in its habitat preference than those just mentioned, although it occurs most frequently on sandy plains and interdune depressions where it constructs extensive and deep burrow systems (Prakash 1975). On the other hand, overlapping distributions are not uncommon in the Rajasthan desert, and species associations have been defined as shown in Table 8.8. It will be evident from this table that while a species may occur in more than one habitat, there is one habitat in which it would seem to do better than in any other, and in which it has the advantage over other species. This illustrates the concept of ecological refuges already referred to in relation to the distribution of tenebrionid beetles in hot deserts.

Habitat selection is not unique to the rodents of the Rajasthan desert. Such selection is also shown by rodents in the deserts of the Middle East, North Africa, and North America (Table 8.9). Schroder and Rosenzweig (1975) and Whitford et al. (1978), for example, carried out manipulative experiments that demonstrated that habitat selection was an important factor in eliminating

TABLE 8.8.
Rodent Associations in Various Habitats in the Rajasthan Desert of India

Habitat	Rodent Associations
Sand dunes	*Gerbillus gleadowi, Meriones hurrianae*
Sandy plain	*Tatera indica, Meriones hurrianae, Gerbillus nanus*
Rocks	*Rattus cutchicus, Mus cervicolor*
Gravel plains	*Meriones hurrianae, Tatera indica*

Note: The first-named species in each habitat is more abundant than the second, etc.

Source: Adapted from Prakash 1975.

interspecific competition between *Dipodomys* species. In particular it was suggested that two congeners, *D. merriami* and *D. ordii*, avoid interspecific encounters by exploiting different parts of the same habitat, with *D. ordii* colonizing grassy sites and *D. merriami* shrub-dominated areas. These two species show considerable overlap in seed-size selection and habitat utilization, according to Brown and Lieberman (1973), but within a particular habitat, *D. merriami* selects more rocky soils and lives in shallower burrows than *D. ordii* (Gaby 1972). As might be expected, *D. ordii* is less able to tolerate high temperatures than *D. merriami*, while the latter is also better equipped for life near the soil surface by having a higher metabolic rate at low ambient temperatures.

Another example, very much along the same lines, is provided by the sympatric pocket mice *Perognathus penicillatus* and *P.*

TABLE 8.9.
Microhabitat Preferences of Some Rodents in the Deserts of the Middle East and North America

Habitat	Gaza Desert, Israel	Nevada Desert
Rock	*Sekeetamys calurus* *Gerbillus dasyurus*	*Neotoma lepida* *Peromyscus crinitus* *Perognathus formosus*
Sand	*Jaculus jaculus* *Gerbillus gerbillus*	*Dipodomys deserti* *Microdipodops pallidus*
Soil	*Meriones crassus* *Gerbillus nanus*	*Perognathus longimembris* *Citellus leucurus* *Peromyscus eremicus* *Dipodomys microps*

Source: Adapted from Eisenberg 1975.

intermedius in the southwestern United States (Hoover, Whitford, and Flavill 1977). Separation of these two congeners is brought about by selection of different soil types in which burrows are constructed. *P. penicillatus* exploits sandy soils in cool, shady sites and probably could exclude *P. intermedius* from such areas by aggression. *P. intermedius* colonizes soils with larger particle sizes; these have a lower heat-buffering capacity and higher burrow CO_2 concentrations. This represents a more extreme environment compared with that experienced by *P. penicillatus*. As may be expected, *P. intermedius* is physiologically better adapted to cope with extreme conditions; it is more tolerant of high and low temperatures than *P. pencillatus*, and it also has a greater ability to restrict evaporative water loss at high ambient temperatures and CO_2 levels.

These preferences for different types of substrates and vegetation types, coupled with different physiological responses, provide mechanisms for coexistence based on spatial separation of congeners. They are reinforced by different patterns of feeding behavior, and we shall be coming to these in a moment. Before we do that, however, it should be noted that the patterns of distribution and diversity that result from the mechanisms of habitat separation discussed above are not static. This point can be documented fairly succinctly.

The stochastic nature of rainfall events in hot deserts means that habitats vary in their suitability for colonization, not only from season to season but also from year to year. As a consequence patterns of rodent species distribution will also vary with time. In the example of the pocket mice, *Perognathus* spp., mentioned above, *P. intermedius* can extend its range during moist periods at the expense of that of *P. penicillatus* but retreats to more mesic conditions at high altitudes during drought (Hoover 1973, quoted in Whitford 1976b). This particular example illustrates the fact that species diversity of rodents within a given site may increase in favorable years owing to the immigration of transient species (Whitford 1976b). In the deserts of the southwestern United States such transients include species of *Peromyscus*, *Reithrodontomys*, and *Mus musculus*. These can establish breeding colonies in deserts when primary productivity is high and competition from heteromyids is low (Brown 1973). Furthermore, these transients can recruit rapidly to achieve high population densities under these conditions since their reproductive potential is higher than that of the resident heteromyids (Whitford 1976b; see also Chapter 3, Table 3.2). During periods of drought (and therefore low primary production) species diversity decreases as the transients disappear, and densities of resident species decline. In considering these events, Whitford (1976b) points to immigration and extinction, linked to changes in primary production, as the causative agents of rapid changes in rodent species diversity and suggests that desert rodent assemblages may be considered as having the attributes of a "habitat island" fauna.

In a very real sense habitat separation must figure prominently in strategies for resource allocation, although, as we will see, this separation is not the whole story as far as desert rodents are concerned. More subtle mechanisms exist. To understand how these operate, it is necessary, first, to review various aspects of desert rodent feeding behavior.

Feeding Behavior. Eisenberg (1975) defined four ecological niches to which desert rodents could be assigned on the basis of feeding behavior:

    Diurnal surface-foraging granivores,
    Nocturnal insectivore-granivores,
    Nocturnal granivore-herbivores, and
    Fossorial types (considered earlier in this chapter).

Diurnal surface-foraging granivores are exemplified by ground squirrels belonging to the genus *Citellus*. Diurnalism is so unusual among desert rodents that the ground squirrels are of particular interest. These rodents forage on the surface, and although Eisenberg (1975) identifies at least some of them as granivores, Hawbecker (1975) and Morton (1979) place them in the omnivore category. To be sure, seeds form an important part of the diet when they are available, but when they are not, other items are taken: green plant material, insects (tenebrionids and crickets) and vertebrates (other rodents and lizards). This plasticity of feeding habit provides a mechanism for diet switching when competition for one particular resource becomes too intense. This may allow for coexistence between potential competitors, although it is essentially a partitioning of resources and, as such, is considered later in this chapter. On the other hand, some of the North American ground squirrels are geographically separated, and this may indicate preferences for different types of habitat. *Citellus mohavensis* is restricted to the northern part of the Californian Mojave desert, while *C. nelsoni* occurs only in the southwestern part of the San Joaquin Valley, a region that has all the characteristics of the Sonoran desert. However, distributional ranges of *Citellus leucurus* and *C. tereticaudus* may overlap in southeastern California, and coexistence may be possible owing to the fact that the former species can utilize seeds as a source of food, while the latter, apparently, does not (Vorhies 1945). Resource partitioning, once again, becomes an important factor in maintaining species diversity.

The nocturnal insectivore-granivore niche is a very specialized one, occupied by the grasshopper mice of the United States, *Onychomys leucogaster* and *O. torridus*. These two species are mainly insectivorous during the warmer months of the year but shift to a diet of seeds when insects are in short supply (Brown, Reichman, and Davidson 1979). They are rarely brought into competitive encounters since *O. leucogaster* inhabits semiarid scrub deserts whereas *O. torridus* lives in more arid regions (Smith and Jorgensen 1975). This ecological grouping differs from the previous one in that it consists of nocturnal foragers, and it is mainly insectivorous, only secondarily granivorous. However, it must be noted that temporal differences in foraging strategies—that is, nocturnalism contrasted with day-active behavior—do not necessarily lead to niche diversification if the food resource remains the same. It does not really matter that two different groups of desert rodents forage at different times during the diel cycle if they are still exploiting the same resource (for example, seeds); they are still competitors. It is the priorities that become important, and herein lies the main difference between this group and the previous one. *Onychomys* spp. are facultative granivores, whereas *Citellus* spp. are facultative insectivores and omnivores.

The third of Eisenberg's groupings, the nocturnal granivore-herbivore niche, is a large one and extends over a number of arid environments including chaparral inhabited by such species as *Neotoma lepida, Peromyscus maniculatus, P. eremicus, Perognathus californicus,* and *Dipodomys agilis*. It is also represented in dry, flat pebble deserts, which are the provenance of *Perognathus formosus, Dipodomys microps,* and *Microdipodops megacephalus,* and in sand deserts where *Perognathus penicillatus, Microdipodops pallidus, Dipodomys merriami,* and *D. deserti* occur. These horizontal differences in species' distribution are reinforced by vertical separation. Rosenzweig, Smigel, and Kraft, (1975) working in the American deserts, distinguished between rodent species that appear to be bush specialists and those that are ground specialists. The former, represented by species of *Peromyscus, Reithrodontomys,* and *Perognathus penicillatus,* climb into shrubs and are as active here as they are on the ground. In contrast kangaroo rats (*Dipodomys* spp.) do not climb. Just how important this vertical separation is in promoting coexistence is something that remains to be established. However, it is worth noting that species of *Reithrodontomys* and *Peromyscus*, which are bush specialists, tend to be omnivores or insectivores in the summer (Brown, Reichman, and Davidson 1979), and they may forage for food in the aerial parts of the vegetation. Kangaroo rats, on the other hand, are seed-feeders and can find their food on the ground.

As already noted, seeds are a readily available resource in hot deserts, and granivorous rodents show different patterns of foraging behavior and resource selection. The different patterns of foraging behavior may serve to minimize interspecific encounters and, thereby, competitive interactions. But they are not absolute since bush specialists switch to a diet of seeds when their preferred food is unavailable. At these times they become potential competitors of the granivorous kangaroo rats. That these two groups of rodents are able to coexist can be attributed to differential resource allocation, and we are going to look at this in more detail now.

Resource Allocation. Habitat preference in itself would not seem to provide a complete explanation for the diversity of rodent species occurring within particular desert regions. Differential allocation of resources is undoubtedly a contributing factor (Rosenzweig, Smigel, and Kraft 1975). The mechanism for allocating resources among granivorous rodents is provided by different foraging strategies, and these have just been discussed. This theme can now be expanded.

The studies of Reichman and Oberstein (1977) on the sympatric kangaroo rat (*Dipodomys merriami*) and the pocket mouse (*Perognathus amplus*) have demonstrated that the former species selects for high resource densities—that is, clusters of seeds—when it forages, whereas *P. amplus* does not show any preference for clumped or dispersed seeds. There is no obvious explanation for these behavioral differences, but their selective value in making coexistence possible is obvious. Again, the bannertail kangaroo rat (*Dipodomys spectabilis*) of North American deserts harvests intact seed heads (Schroder 1979), whereas other species of this genus seek out discharged seeds. This may provide yet another example of resource allocation.

Dipodomys spp. are bipedal and have a saltatory method of locomotion, which allows them to locate clusters of seeds more efficiently than the quadripedal pocket mice (*Perognathus*). The latter are random feeders, sifting through large quantities of soil and litter to collect seeds (see Brown, Reichman, and Davidson

1979). Moreover, as we have already seen, such species as *Perognathus penicillatus* are bush specialists and may exploit seed reservoirs unavailable to ground-active *Dipodomys*.

In all probability coexistence among desert rodents is achieved by virtue of a combination of behavioral and ecological adaptations. To illustrate this we can shift our attention from the deserts of North America to those of the northern Sahara—more specifically to the northern parts of the Sudan where the desert is sandy or stony. This is the home of five common desert rodents, the sciurid ground squirrel (*Euxerus erythropus*), the dipodid jerboa (*Jaculus jaculus*), the cricetid gerbils (*Gerbillus pyramidum* and *G. watersi*), and the sand rat (*Meriones crassus*). Two of these species, *Jaculus jaculus* and *Gerbillus pyramidum*, are relatively common, and they are sympatric and show a nocturnal activity pattern (Happold 1975). They coexist because they have contrasting modes of behavior and ecologies. The jerboa (*Jaculus jaculus*) is solitary, moves away from the burrow at night, does no eat hard seeds or fruits, and probably does not store food. The gerbil (*Gerbillus pyramidum*) tends to be colonial, will accept hard seeds as part of the diet, and lays down large food stores. In passing, it is interesting to recall that two North American desert scorpions that live sympatrically have identical mechanisms for resource allocation. *Diplocentrus spitzeri*, like *Gerbillus pyramidum*, forages around the burrow entrance, whereas *Centruroides sculpturatus* and *Jaculus jaculus* forage away from their burrows.

From these kinds of comparative observations there emerges a central idea that can lead to an understanding of how species diversity of surface-active foragers can be maintained in hot deserts. The strategies for survival are limited, and each of these strategies has apparently been discovered independently by taxonomically unrelated groups. We have seen that there are parallels between various rodents and ants in the way in which foraging behavior is related to resource density. Rodents and scorpions both show spatial differences in their foraging strategies that contribute to coexistence of related species. Finally, some measure of habitat specificity is shown by both rodents and tenebrionid beetles, and as a consequence ecological refuges permit some measure of species diversity to exist.

These parallels also indicate the very real possibility, among granivores at least, of competition for food between taxonomically unrelated groups, such as the rodents and the ants. This is unexpected since evolutionary theory would lead us to believe that competition should be more intense between taxonomically related groups than between those more remote in this respect. However,

Brown and Davidson (1977) found that the density of the harvester ant *Pheidole* sp. increased in areas where seed-feeding rodents were excluded, suggesting that these seed harvesters responded quickly to the removal of competitors (Whitford et al. 1981). This work also showed, however, that harvester ants of the genus *Pogonomyrmex* were not affected by the removal of rodent competitors. Evidently, some form of stable coexistence has evolved between these competing classes that Brown, Reichman, and Davidson (1979) explain in the following ways.

1. Ants appear to harvest seed-size classes in proportion to their occurrence in the environment. They take smaller seeds, on average, than rodents, which seem to forage selectively for large seeds.
2. Ants are better able to cope than rodents with the unpredictability of resources through their regulation of colony size and foraging activity.
3. Ants are able to gain greater rewards by foraging on lower densities of seeds than are rodents, owing to the smaller body size of the worker caste of the former group.
4. Rodents are more efficient than ants in collecting dense accumulations of seeds, and they are more efficient in digging for buried seeds.

The importance of each of these factors may well vary from one region of the desert to another.

9
Life-Styles for Survival: Tactics

Much of what has gone before in this book has been concerned with the ways in which animals, particularly those associated with the soil, have developed, or taken advantage of, life-styles suited to a desert existence. These life-styles are variations on a few major themes that combine in different ways to reduce exposure to environmental stress or provide the means to tolerate this stress. As this book draws near to its conclusion, it may be useful to highlight the features of these life-styles. In doing so it is convenient to distinguish between individual traits (tactics) and the constellation of traits (the strategy) that characterize the complete life-style. This chapter is concerned with tactics; the following one with strategies. Before we embark on this synthesis, however, it is necessary to clarify what is meant by *environmental stress*, as it is used here.

In hot deserts extremes of such environmental variables as temperature and moisture can produce physiological stress in the animals and plants living there. As far as the temperature factor is concerned, stress occurs when the body temperature exceeds the upper and lower threshold limits of tolerance. Body temperatures in excess of the upper tolerance limit occur when heat is accumulating faster than it can be dissipated, and the outcome is explosive heat death (Schmidt-Nielsen 1964). At the other end of the scale a decrease in body temperature below the lower tolerance level results in a retardation of the metabolic rate to a point at which the energy demands of cells, tissues, and organs cannot be satisfied.

Further, if body temperatures fall to zero or below, water is rendered unavailable by the formation of ice crystals. Not only does this result in a concentration of body fluids, but crystal formation may cause structural damage to the cytoplasm. The concentration of body fluids may also occur when the environmental factors of high temperature and low moisture availability combine, particularly in those animals that employ evaporative cooling as a means of regulating body temperature. Increased loss of water from the body at high temperatures through transpiration will increase the viscosity of body fluids, notably the blood, and the role of these fluids as a transport system will be gravely impaired. Particularly important in this respect is the reduced conduction of heat from the body core to the periphery, where it can be dissipated to the environment by convection and conduction; this is an important factor in the accumulation of body heat mentioned above. On the other hand, excess moisture in the environment can also produce physiological stress in animals that cannot control water uptake. This has the consequence of diluting body fluids to a point at which attenuation of metabolic activities results in "water intoxication" (Schmidt-Nielsen 1964).

Expressed in these terms, the idea of physiological stress is unambiguous if tolerance thresholds, with respect to each environmental variable, can be identified accurately and precisely for a group of individuals living in the same locality. Techniques are available to monitor body temperature changes, transpiration rates, blood volume and osmolality, tissue fluid levels, and respiratory substrates. The use of these techniques in conjunction with visual observations on changes in the behavior pattern of an animal (or animals) exposed to a gradient of an environmental factor will serve to identify tolerance limits. There is no doubting, for example, when a kangaroo rat (*Dipodomys* sp.) is undergoing temperature stress, for it salivates over the fur on its throat region; it cannot sweat, and so it adopts this method of evaporative cooling. Desert tortoises urinate over their posterior limbs to produce the same effect. The locomotion patterns of arthropods become uncoordinated at temperatures above tolerance thresholds, and when these animals are intolerably chilled, they become comatose.

It must be remembered, however, that tolerance limits are not immutable; they are very much a function of conditioning of an animal or a localized group of animals. As such, they may vary temporally and spatially even within the same species, and with them, of course, will vary the stress thresholds. If this is not understood, generalizations about these thresholds will necessarily be imprecise and ambiguous.

Avoidance and tolerance are the two general strategies that we have identified with the survival and success of desert soil animals. There is not always a hard and fast distinction between the two, and many desert soil animals have a life-style that incorporates elements of both of these strategies. It is perhaps more perspicuous to examine these life-styles in terms of morphological, behavioral, and physiological tactics.

MORPHOLOGICAL TRAITS

The adaptations that can be considered under this heading involve structural features that are associated in part with avoidance, in part with tolerance. They are dealt with under three main headings.

Size and Shape

A relatively large body size, with the consequent reduction in surface area/volume ratio, is obviously advantageous to an animal living in hot deserts. Transpirational losses across the body surface and heat gain from the environment are reduced to a minimum in such large-sized forms, and it comes as no surprise, therefore, that desert centipedes, millipedes, and tenebrionid beetles, for example, are larger in size than their moist temperate relatives. What is perhaps surprising is that many groups of the desert soil fauna do not conform to this trend. The desert snails, *Sphincterochila zonata* and *Trochoidea seetzeni*, are no larger than many of their moist temperate relatives and, indeed, are considerably smaller than some. Desert scorpions, particularly those that live in the more arid regions, tend to be larger in body size than their mesic relatives, but there are exceptions. Similarly, desert woodlice are not markedly larger than some of the woodlice found in cool, moist temperate forests.

These observations indicate that there is no universality in selection for large body size among desert soil fauna; while this particular morphological adaptation may be useful in certain cases, it is not an essential feature of the life-style of these animals. The same conclusion may be applied with equal force to the adaptive features associated with body shape.

The minimum exposure of body surface to the impact of environmental factors (desiccation and temperature) is achieved by animals that can assume or present a spherical body shape. Desert tenebrionids have a subcylindrical shape, in contrast to their dorsoventrally flattened mesic relatives, and this can be interpreted as a device for minimizing transpirational losses and uncontrollable temperature gains to and from the external environment. The desert woodlice *Venezillo arizonicus*, *Armadillo officinalis*, and *A. albomarginatus* have the ability to conglobate and, thereby, assume a spherical shape, which effectively seals off their respiratory surfaces (see Chapter 6). The spiral coiling of the desert millipede *Orthoporus ornatus* is designed to produce similar results. On the other hand, the assumption of, or approximation to, a spherical shape is not universal among desert soil animals. The sand roach, *Arenivaga*, is markedly flattened dorsoventrally, as are the scolopendromorph centipedes and the polydesmoid millipedes. This kind of body shape is advantageous for a burrowing animal and is associated with a life-style that places a premium on the avoidance of environmental stress. Again, one of the most successful groups of desert woodlice, *Hemilepistus* spp., cannot conglobate and relies for its survival on its ability to burrow. These isopods are subcylindrical in shape, as are the burrowing rodents. These burrowers excavate tunnels in the soil that have a measure of permanence, in contrast to the centipedes and millipedes that often seek the shelter of preexisting crevices.

Coloration

The adaptive significance of body coloration in desert animals is open to various interpretations, and the review by Hamilton (1975) provides a particularly good introduction to the current state of the art. Perhaps a good place to start is to identify the areas in which there is a measure of agreement.

In the first instance coloration of desert animals can subserve one or more of three functions, at least: communication, concealment, and optimization of radiation relationships. The ability of animals to communicate rapidly and effectively with each other is likely to be of considerable importance in systems where a premium is placed on cooperation rather than competition. Activity on the soil surface, particularly during the daytime, renders an animal vulnerable to predators, and this vulnerability will be enhanced in deserts where plant cover is discontinuous (see Fig. 8.3). Concealing coloration (see Fig. 3.3) is clearly advantageous in such circumstances. Poikilothermous animals need to utilize the thermal characteristics of their environment in the most efficient way if their activities are to be economic. To do this they employ behavioral and morphological adaptations. The former, which include burrowing and orientation behavior, will be considered in a moment. Morphological traits associated with thermoregulation essentially involve coloration, and these are considered here.

Second, shiny and highly reflective body surfaces are perhaps the exception rather than the rule in hot deserts. Tenebrionids of the genera *Eleodes* and *Cryptoglossa* are moderately reflective in their general appearance, to be sure, but they pale into insignificance when compared with mesic carabids such as *Nebria brevicollis* (Figs. 9.1 and 9.2). Desert snails of the genus *Sphincterochila* do not possess the highly polished appearance of the mesic *Cepaea nemoralis*, and this lends support to the idea that dull, rather than

Fig. 9.1. Tenebrionids from the Mojave desert, California. Upper specimen is showing aggressive posture; lower one is faking death. (Photo by G. Ott.)

Fig. 9.2. The carabid beetle *Nebria brevicollis* from beech woodland, England. Note the highly reflective body surface. (Photo by Haidee Price-Thomas.)

shiny, surfaces have a selective value for desert soil animals. Hamilton (1975) suggested that the advantage of such dull coloration lies in its camouflage properties and has little to do therefore with thermal relationships. This is borne out by the finding, reported in Chapter 7, that despite its dull appearance the shell of *Sphincterochila zonata* reflects about 90 percent of the incident radiation falling on it.

Third, the reflection of incident radiation is very much a function of color (or the lack of it, to be very precise), and black and white are of overriding significance; other colors are subsidiary and are deployed for other purposes—communication and concealment. White surfaces are more reflective than black ones under all ambient conditions (Hamilton 1975); indeed, we have already noted in Chapter 7 that thermal loading is less in tenebrionid beetles with white elytra (such as *Onymacris bicolor* and *O. candidipennis*) than in uniformly black species (*Onymacris plana, O. laeviceps,* and *O. unguicularis*) (Hadley 1970a; Hamilton 1975). However, our interpretation of Hadley's results differs from that of Hamilton since although we recognize body temperature differences between black and white forms, we do not consider these to be significant, whereas Hamilton does. However, Hamilton points out that solar radiation is not transmitted through black cuticles, and any energy passing through white cuticles is absorbed by the brown lining of the underside of the elytra. To this extent, then, it is immaterial whether the body covering is black or white, a conclusion already reached in Chapter 7, albeit by a different line of reasoning that invokes the buffering effect of the subelytral air space.

To this point we have been particularly concerned with the protective function of coloration, with its role in reducing thermal loading. However, perhaps this is the wrong approach. The mobility of day-active lizards and various tenebrionid beetles, coupled with their ability to burrow or seek shelter in preexisting soil refuges, probably prevents these animals from undergoing thermal stress during periods of sunlight. Undoubtedly, black beetles experience heat gain in direct sunlight—more so than the white forms—but, as Hamilton (1975) argues, this can be viewed as a positive advantage rather than a hazard. Desert tenebrionids tend to maintain an elevated body temperature (between 38°C and 40°C) for as long as possible; the advantages of this are seen as a rapid conversion of food to energy and an accelerated rate of growth and reproduction. If the body temperature falls below 38°C, behavioral mechanisms are invoked to maximize heat gain from the environment, and to this end black coloration is a positive asset. Indeed, any adaptations that enhance heat gain and heat retention will have selective value for desert animals living in conditions subject to large circadian and seasonal fluctuations in temperature. Such adaptations may reduce the risk of cold torpor during the night, and this will apply both to black and white forms. White has the slight advantage over black during the daytime since reduced thermal loading will permit more extended periods of surface foraging activity. This point is underlined by the fact that desert lizards undergo color changes as the body temperature rises. *Dipsosaurus dorsalis,* for example, assumes the "dark" phase at temperatures between 33°C and 38°C but blanches when the body temperature rises to 40°C and above (Norris 1967). Similarly, the tenebrionid *Stenocara phalangium* exhibits immediate and seasonal color changes in response to changing temperature conditions (Hamilton 1975).

Insulation

The development of insulating cushions of air, partly or completely surrounding the body, has occurred in at least three completely unrelated groups of desert soil animals: snails, tenebrionid beetles, and rodents. The snail *Sphincterochila zonata* has a disproportionately large, air-filled lower shell whorl, which reduces heat flow from the exposed rock surfaces on which the snail lives to the body. The subelytral space of desert tenebrionids performs the same function. Again, surface-active desert rodents, in contrast to some of their more completely subterranean relatives, have a thick pelage, which effectively traps an insulating layer of air around the body.

The morphological adaptations that we have just been considering illustrate evolutionary convergence between unrelated groups of the desert soil fauna. Examples of this kind of convergence are limited: many desert soil animals are no larger in body size than their mesic relatives, for example; some of these animals do not have highly reflective body surfaces, and, indeed, relatively few can employ the insulating effects of air cushions to minimize thermal stress. The animals that have been able to put these adaptions to good effect in hot deserts are those that are structurally preadapted and that have been able to capitalize on this in what can be regarded as an opportunistic manner. For instance, there is a general tendency among tenebrionid beetles in mesic environments for the elytra to be fused together. This feature is a prerequisite for a subelytral space to operate as an efficient thermal insulator. Again, the spiral design of the shell of a pulmonate gastropod mollusk is such that the basal whorl is always the largest, and it requires little differential growth to ensure that this whorl becomes large enough to accommodate an effective insulating air cushion.

Moving on now to the second category of survival tactics, behavioral adaptations, examples of convergent evolution become more common.

BEHAVIORAL TRAITS

The adaptations considered in this section are designed, for the most part, to remove desert soil animals from situations of potential stress. The particular stressed situations with which we are concerned are extremes of temperature and moisture and scarcity of food.

Circadian Rhythms and Orientation Behavior

The list of desert soil animals that are nocturnal in habit is a long one: scorpions, solifugids, uropygids, centipedes, woodlice and many ants, termites, tenebrionid beetles, and rodents. These have been discussed in the context of this behavior pattern in previous chapters, and there is no necessity to go over this ground again. Suffice it to say that nocturnalism is an adaptation that operates to minimize water loss to the environment, maximize water gain from the environment, allow avoidance of excessively high surface temperatures, and reduce predation risks. Again, as we have already seen, nocturnal activity patterns are not inflexible but may be subject to seasonal variation. This flexibility could present difficulties in defining a truly nocturnal desert animal. This dilemma can be resolved, in some cases at least, by identifying anatomical features that have a selective value for an animal that is primarily nocturnal. One example will serve to illustrate this point. This is provided by the nocturnal-foraging heteromyid rodents of North American deserts, particularly members of the genera *Dipodomys* and *Microdipodops*. These surface-active granivores are prone to predation, and enhanced visual acuity is of no real benefit during the hours of darkness. However, increased auditory sensitivity provides early warning of the approach of a predator. This sensitivity is mediated through a greatly inflated middle ear, a very efficient tympano-ossicular system, and an ability to detect low-frequency sounds (see Chapter 3). This is a good example of the way in which a morphological feature (modification of the ear) and a behavioral tactic (nocturnalism) combine to produce a life-style for survival.

Desert lizards, millipedes, and snails use orientation behavior as an aid to thermoregulation. By moving vertically on rock faces (for example, snails of the genus *Cristataria*) or laterally through shrub canopies (millipedes and lizards), these animals can position themselves in temperature gradients so as to utilize these for metabolic purposes in the most efficient way. This tactic has been discussed in some detail in Chapter 7.

Burrowing Behavior and Sociality

The ability to burrow in the soil is so widespread among desert animals that this fact in itself reflects the vital importance that can be attached to this ability in an environment where escape from the soil surface may spell the difference between life and death. As we have seen on many occasions in the preceding chapters, there are many morphological, behavioral, and physiological adaptations associated with the burrowing habit. It has also become evident that animals differ in their burrowing ability and that burrows vary in their degree of permanence.

Many of the animals that produce relatively permanent burrow systems (various scorpions, *Hemilepistus*, ants, termites, tortoises, and rodents) exhibit sociality, and the focal point of social organization is the burrow system. Ants and termites, with their elaborate social structure and caste differentiation, are prime examples of this type of organization. Colony formation also occurs, as already noted in Chapter 4, among such subterranean mammals as the South American tuco-tuco and octodont, the South African mole rat, and the naked mole rat of North Africa when food supplies are plentiful. Social organization of a rather different kind is shown by such strange bedfellows as the scorpions and the wood louse *Hemilepistus*—namely, the family group. This has already been discussed in the appropriate places in Chapter 2, and it will be mentioned again in the context of reproductive tactics later. For the moment it is necessary only to make the point that scorpions may be particularly vulnerable to predators when the mother is encumbered with her young, and access to a soil burrow will provide some measure of protection. In this event the young benefit from their association with the parent. The case of *Hemilepistus* is rather different. The members of this woodlouse genus are unique among terrestrial isopods in that the young do not disperse after liberation from the brood pouch. Instead, they remain in the parental burrow and actively participate in the vertical excavation of the burrow system. Without this participation it is unlikely that the burrow could attain the depth required for the survival of the young until the next rains (see Chapter 2).

Seasonality

The model of desert ecosystem functioning presented in Chapter 5 postulates discontinuous biological activity that is triggered by aseasonal or seasonal rainfall events. This postulate can now be examined in more detail with particular reference to two kinds of biological activity: feeding and reproduction.

Some groups of desert soil fauna are active only for a very short period of time each year, which coincides with the rainy season. In the case of the millipede *Orthoporus ornatus* this active period lasts for three to four months. The snails *Sphincterochila zonata* and *S. prophetarum* are active for less than a month each year (Shachak, Chapman, and Steinberger 1976; Yossi Steinberger 1980: personal communication), while spadefoot toads of the genus *Scaphiopus* may be active for less than a week in a 12-month period. During these relatively short periods of time, feeding, mating, and egg laying occur. These are, perhaps, extreme examples of restricted activity, and they refer to animals that live in deserts characterized by short or irregular rainfall events. As one might expect, the duration of biological activity, considered on an annual scale, will be related to the duration of the rainy season, and this will vary from desert to desert, particularly for those groups that are habitually diurnal. Nocturnalism may allow a species to extend its activities into the dry season, but most desert soil animals estivate during the summer and/or hibernate during the winter, at least for short periods of time. Many examples of these have been cited in the previous chapters.

Feeding Behavior. *Orthoporus ornatus*, *Sphincterochila* spp., and *Scaphiopus* spp. are decidely discontinuous feeders and, as such, conform very nicely to the pulse-reserve model mentioned earlier. Their ability to convert food into body fat very rapidly lies at the heart of this particular tactic. Indeed, *Scaphiopus couchii*, feeding on alate termites, can consume 55 percent of its body weight in one feeding and, as a result, store enough energy reserves to last for 12 months (Dimmitt and Ruibal 1980b). *Orthoporus ornatus* has an assimilation efficiency of 20 to 36 percent (Nunez and Crawford 1976; see also Chapter 2), which is remarkably high for a terrestrial detritivore.

However, these millipedes, snails, and anurans are not necessarily typical of desert soil animals in general. To be sure, the organic base for desert soil ecosystems has a discontinuous input that occurs when annual plants undergo vegetative growth, flowering, and seed production, followed closely by seed discharge and dieback. Leaf fall from perennial shrubs is also seasonal. This plant material that falls on the soil surface may persist for long periods of time and provide a continuing food resource for detritivores and granivores. The feeding activities of these trophic groups are not, then, subject to seasonal constraints on food supplies. Cryptostigmatid mites living in relatively permanent accumulations of leaf litter in organic soils may be stimulated into activity by the presence of environmental moisture occasioned by dew formation. This is a circadian phenomenon that may persist throughout most of the year. Another tactic that serves to extend the period of feeding activity is that of diet switching.

The ability to vary the diet—in effect to switch from one food resource to another, depending on the immediate availability—carries with it a great deal of ecological flexibility. Not only does it allow for an extension of period of feeding and food accumulation, but it may also reduce the detrimental interaction between two or more species competing for the same diminishing resource. African desert rodents of the genus *Meriones* will feed on the green aerial parts of annual plants when these are available during the summer season but retire into a granivorous feeding refuge at other times of the year (Chapter 3). During this granivorous period they are brought into competitive situations with other granivorous rodents and ants; the mechanisms whereby these encounters are minimized have already been reviewed in Chapter 8. Diet switching is also a feature of the life-style of such ants as *Novomessor* spp. These are mainly granivorous but will prey on other arthropods if the occasion demands. Again, ants of the genus *Messor* in Mediterranean deserts switch from feeding on seeds of *Jasonia* in February to seeds of *Elymus* in May (Ghabbour, Mikhail, and Rizk 1977). Flexibility of feeding habit is also demonstrated by the horned lizard, *Phrynosoma modestum*, which feeds mainly on honey-pot ants of the genus *Myrmecocystus* but will switch to other prey items such as *Pheidole* spp. and *Conomyrma* spp. during the rainy season when these are available (Schaffer and Whitford 1981).

Diet switching in the cricetid *Peromyscus crinitus* was detected by MacMillen and Christopher (1975) from analyses of urine samples taken in the field. Seasonal variations in urine concentrations allowed these authors to conclude that this rodent was carnivorous (probably insectivorous) in winter but possibly granivorous during the summer. *P. crinitus* is approaching a state of water independence but retains the ability to tap water sources in the body of its prey, as the occasion demands.

These are just a few examples of diet switching among unrelated groups of the desert soil fauna. It is probable that there are many more as yet undescribed since survival in extreme environments often depends on cooperation between different groups of organisms. In this context cooperation expresses itself by minimizing interspecific competition, and this is just what diet switching is designed to do.

But there is another tactic, admittedly more limited in its scope, for minimizing food stress, and this is food storage. Many granivorous rodents hoard seeds in subterranean larders or caches. These are food supplies set aside to carry the animals through periods of food shortage. Honey-pot ants, which store their liquid food in the bodies of "repletes," adopt the same strategy. Termites of the family Macrotermitidae cultivate fungus gardens for the same purpose. This practice of food storage is perhaps more strongly developed in animals that do not resort to diet switching as a survival strategy. For example, kangaroo rats (*Dipodomys* spp.) feed more or less exclusively on seeds and show caching behavior. On the other hand, *Meriones* spp. show diet switching and only hoard food supplies erratically.

Reproductive Behavior. That reproductive activity is geared to coincide with the rainy season in deserts has been a recurring theme throughout the previous chapters. It is illustrated by such surface-active groups as millipedes, snails, woodlice, and amphibians, and there is some evidence to suggest that many of the more truly subterranean soil arthropods also show this seasonality (see Chapter 4). However, there are variations on this theme, as we have seen. First, there is the delayed reproductive response, and two examples of this have been cited earlier. The scorpion *Paruroctonus mesaensis* breeds during the summer months in the California Mojave desert, a desert that receives most of its rainfall in winter. The reason for this may lie in the fact that mating occurs in the spring, but the young are retained within the uterus of the parent until their development is well advanced. During this time the mother nourishes the embryos from the food she has been able to obtain from prey captured in the spring, in all probability (Polis and Farley 1979). In this same desert region the soil-dwelling cryptostigmatid mite *Haplochthonius variabilis* recruits a new generation in the period April to June, some three to four months after the onset of the winter rains. In this case the higher spring temperatures may be the trigger that sets off the breeding response (see Chapter 7).

These two examples are really variations on the seasonal theme, for in both cases breeding occurs during a restricted period of the year, which, at least in the scorpion *P. mesaensis* appears to be fixed from one year to the next. This built-in reproductive response may be a reflection of the fact that this predatory arachnid lives in a desert in which the rainfall event is predictable, at least on a seasonal basis. However, the rodents of North American deserts present a rather different picture and illustrate a second type of exception to the seasonality phenomenon. Many of these granivores are polyestrous and can breed throughout the year. This is true of members of the genera *Onychomys*, *Peromyscus*, *Reithrodontomys*, and *Dipodomys*, although some members of the last genus may have restricted breeding periods (Chapter 3). Polyestrous also occurs in desert marsupials, and this has been interpreted as an opportunistic adaptation to life in an environment where rainfall is unpredictable and aseasonal (Tyndale-Biscoe 1973). However, many small rodents living in more stable and predictable cool moist temperate habitats are also polyestrous. Desert marsupials, in contrast to true rodents, have a relatively high reproductive output (Chapter 3), and this may be a device to compensate for high levels of juvenile mortality. Among the rodents the Australian murids live in deserts where the rainfall is extremely variable; these rodents have larger litter sizes than most, and life expectancy is extremely short (see Table 3.2). Similarly, the woodlice *Hemilepistus reaumuri* and *Armadillo officinalis* have brood sizes of 80 to 90 and 45, respectively, and these are substantially higher than that of the semiarid *Porcellio olivieri* with 8 to 32 (Shachak 1980). It will be recalled (Chapter 2) that the last-named species can produce more than one brood in a season, and this would also seem to be the case with *Metaponorthus pruinosus* and *Leptotrichus naupliensis* (El-Kifl et al. 1970). In contrast *H. reaumuri* and *A. officinalis* are single brooded (El-Kifl et al. 1970; Shachak 1980). From the review presented in Chapter 2 it may be concluded that relatively short life-spans are probably the rule rather than the exception among woodlice of arid and semiarid regions. However, it would be a mistake to assume that all desert isopods follow the example of *H. reaumuri* and *A. officinalis* in producing large broods. Warburg (1965b) reported that *Venezillo arizonicus* and *Armadillidium vulgare* in the Sonoran desert have smaller clutch sizes than mesic populations of *A. vulgare*. This fact immediately alerts us to the possibility that more than one option, in terms of reproductive tactic, may be open to desert soil animals.

It may be argued that animals living in unpredictable environments will be under selective pressure to produce large numbers of young to counteract high levels of mortality. This particular tactic, however, is costly from an energetic point of view since there is a low return on the energy invested in reproduction. Further, this investment will be large over a short time period, and often it results in wildly fluctuating population densities. One example of this is the xerophilic woodlouse *Periscyphis granai* studied by Kheirallah (1979) in the western highlands of the Arabian desert. This species has a restricted breeding season (January to March), it has a life-span of about three years, and it probably breeds twice during this time. The recruitment data provided by Kheirallah are not easy to interpret, but it would appear that brood size is relatively large (40 to 50 young per pouch, after correcting for pouch mortality). Survival to the end of the first year of development is only about 4 percent, and population fluctuations within a year are markedly greater than in mesic species.

Another example is provided by the murid rodents of Australian deserts, which can vary in time from plague proportions to virtual extinction. One obvious alternative to this is to reduce energy expenditure by reducing the number of young produced. This appears to be the tactic adopted, not only by the woodlice *V. arizonicus* and *A. vulgare* just mentioned but also by heteromyid rodents (see Table 3.2). A comparison of the data for arid and semiarid millipedes (Table 9.1) suggests that *O. ornatus* belongs in this category, although the absence of data on the number of clutches per season prevents a more meaningful comparison (Clifford Crawford 1980: personal communication).

The production of a reduced clutch size implies that only a low replacement rate is required to compensate for juvenile mortality. Immediately, we can inquire why only a low replacement rate may be appropriate in the extreme environments provided by hot deserts. Two possibilities at least suggest themselves. First, an adult female may compensate for the production of a small clutch by breeding more than once during her reproductive life. By doing this it can spread its reproductive effort, and the energy that is channeled in this direction, over an extended period of time. This has the advantage not only of reducing energy costs at any particular moment in time but also of ensuring the survival of some broods, even though others may be exterminated by unfavorable environmental conditions. In current ecological parlance this is termed bet-

TABLE 9.1
Clutch Sizes of Four Species of Arid and Semiarid Zone Millipedes

Species	Habitat	Mean Clutch Size	Range
Orthoporus ornatus[a]	Arid	500	Not known
Spirostreptus assiniensis[b]	Semiarid	731	456–1,034
Oxydesmus sp.[b]	Semiarid	1,287	631–1,776
Habrodesmus falx[b]	Semiarid	1,436	576–2,353

[a] Data from Crawford and Matlack 1979.
[b] Data from Toye 1967.
Source: Compiled by the author.

hedging. Second, juvenile mortality may be minimized by parental protection of the young.

What seems to be emerging from this discussion is that there is no single generalized reproductive model that can be applied to desert soil animals. Longevity, clutch size, number of clutches produced in a lifetime, duration of gestation, rate of postembryonic development, and survival of the young are all features that express themselves in one way or another through reproductive behavior. The reality is complicated.

In attempting to unravel this complexity a distinction can be made immediately between species that are relatively short-lived—that is, have a life-span of little more than a year at most—and those that live for several years. To the former category belong species that produce only one generation of immatures during the annual cycle. These are termed univoltine and are exemplified by *Hemilepistus*, solifugids, certain spiders, and possibly scolopendromorph centipedes and the cryptostigmatid mites *Joshuella striata* and *Haplochthonius variabilis* (see Chapters 4 and 5). Also included here are species in which postembryonic development is rapid enough to allow several reproductive generations to follow each other in succession within the space of a year. These species are multivoltine, and perhaps it is appropriate to include under this heading *Porcellio olivieri*, which may be classified more strictly as bivoltine (Chapter 2). We might also expect desert soil nematodes and prostigmatid mites to be multivoltine, if the reproductive behavior of their mesic relatives is anything to go by. However, the reproductive biology of desert-dwelling members of these groups is virtually unknown.

Examples of multivoltine groups are likely to be less common among that element of the desert soil fauna that is surface active than among the completely subterranean fauna. In general the former have an activity pattern that is discontinuous and rather sharply restricted to a short rainy season. There would seem to be little scope, under these constraints, for the evolution of multiple generations within an annual cycle. By contrast those groups that can live entirely within the desert soil may not have such restrictions placed on their activity. The life-styles portrayed by this element of the desert soil fauna place emphasis on the avoidance of random and unpredictable events occurring at the soil surface. The animals in this category are avoiders, rather than tolerators, and by living in an equable environment can perhaps indulge in the luxury of the multivoltine mode of life. One of the characteristic features of this life-style is parthenogenetic reproduction (Stearns

1976), and it is worth recalling that a species of cryptostigmatid mite, *Haplochthonius simplex*, inhabiting Mediterranean regions probably reproduces parthenogenetically (see Chapter 4). The related *H. variabilis*, which ranges widely through the deserts of the southwestern United States, has been considered to be univoltine, but the evidence for this is based on observations conducted through less than one annual cycle (Wallwork 1972a). A more detailed examination of the reproductive biology of this species— and indeed of the subterranean fauna of hot deserts in general—is required before the precise nature of survival strategies can be identified. Again, one must not lose sight of the fact that some subterranean groups at least respond to the events occurring on the soil surface and may go into quiescent states during periods of drought; anhydrobiosis in nematodes and similar states in Collembola are illustrative of this (see Chapter 4).

The surface-active component of the desert soil fauna contains a number of groups that are relatively long-lived. These include the herbivorous and detritivorous snails, millipedes, ants, termites, thysanurans, and many, but not all, of the beetles belonging to the family Tenebrionidae. They also include carnivorous arachnids, such as the scorpions and large spiders, and possibly some of the desertic scolopendromorph centipedes, although virtually no information is available on the reproductive biology of the latter. All of these groups have relatively long lives as adults, but we must also include in the long-lived category groups that may be short-lived as adults but take several years to grow to maturity. It is important to identify this category since it comprises many insects that may be active for relatively long periods of time as larvae in the desert soil but that have only an ephemeral existence as adults in aerial environments. Personal observations in the Chihuahuan desert of New Mexico indicate that dipteran and coleopteran larvae are consistently present in organic soils; from what is known of the biology of these two insect groups in mesic soils, it is reasonable to assume that a considerable portion of the life cycle is spent in the juvenile stage. The short-lived adult has only limited opportunities for reproduction and, indeed, may only breed once during its lifetime; this is termed semelparity, and intuitively one would expect species adopting this particular reproductive tactic to produce large numbers of young. Data to substantiate this are largely lacking, although it will be recalled that the semelparous *Hemilepistus reaumuri* and *Armadillo officinalis* have substantially larger clutch sizes than *Porcellio olivieri*, which is certainly bivoltine and may be iteroparous.

The extension of the adult life-span over several years opens up the possibility for repeated breeding; this is termed *iteroparity*, and it obviates the need for large clutch sizes. The millipede *Orthoporus ornatus* is evidently long-lived and iteroparous (Crawford 1976) and, as we have seen, would appear to have a clutch size that, on average, is substantially lower than that of semiarid millipedes. However, Clifford Crawford (1980: personal communication) has found egg production in the Middle Eastern *Archistreptus* sp. to be considerable, and if this millipede should prove to be long-lived, then the line of argument that suggests that iteroparity is closely linked with small clutch size comes into question. Be this as it may, the sand scorpion *Paruroctonus mesaensis* of the southwestern United States can live for at least five years after attaining maturity (Polis 1979), and here iteroparity is certainly associated with a relatively small clutch size (see Table 2.1). Another American desert scorpion, *Centruroides sculpturatus*, may have adopted a similar reproductive tactic, for it produces only about 20 juveniles per clutch (Williams 1969) and is iteroparous to the extent that it can produce at least two clutches per year; the life-span of the adult of this scorpion is, however, unknown.

Scorpions can be considered as some of the best adapted of desert soil invertebrates. They are nocturnal, the majority can burrow in the soil, and they have low basic transpiration rates (see Chapter 6). However, when we try to apply these kinds of generalities to reproductive tactics, the logical approach is found to be wanting. For example, the African *Androctonus australis*, *Leiurus quinquestriatus*, and *Buthus occitanus* produce larger clutch sizes than *P. mesaensis* and *C. sculpturatus* (Chapter 2) and have relatively short gestation periods. All of these species reach maturity within the space of two years (less than one year in the case of *P. mesaensis* and *A. australis*), and it may be presumed that they are iteroparous as a rule. In contrast the Australian *Urodacus yaschenkoi* appears to be semelparous even though it can live for six years (Shorthouse 1971). This oddity can perhaps be explained by the fact that *U. yaschenkoi* takes over four years to reach maturity and lives as an adult only for about 18 months. Inexplicably, however, this semelparous species has a mean clutch size (11.8) that ranks among the smallest of any scorpion.

In view of these remarks, and those made earlier in Chapter 2, it is very difficult to discern any common pattern in the reproductive tactics of desert scorpions. Clutch size and number, gestation period, age at maturity, and the length of adult life show a considerable amount of variation from one species to another and, in some cases, within a particular species (see the data compiled by Polis and Farley 1979). Again, the time at which matings and births occur is very variable. The North African *Leiurus quinquestriatus* can mate and give birth all the year round, whereas the North American *Paruroctonus mesaensis* mates from May to October and births occur in synchrony the following August.

This variability may well be a reflection of the variability and, indeed, unpredictability of food resources. Prey abundance, in turn, depends on climate, particularly the amount of precipitation and the annual temperature range, and on the amount of vegetation available to detritivores and herbivores. These environmental factors vary considerably in their expression from one desert to another, and within any given desert region, as we have seen. When the reproductive biology of desert scorpions is placed in this context, its variability becomes more understandable, although some inconsistencies still remain. One of these relates to parental care.

Young scorpions are brooded on the back of the female parent until they are ready to molt into the second instar. After this time they leave the body of the parent but may remain in the vicinity of the breeding site for some time. Parental care is clearly a device dedicated to the maximum survival of the offspring, at least in theory. In the case of *Paruroctonus mesaensis*, however, Polis and Farley (1979) found that only 40 percent of embryos survived to the second instar, and this high mortality could be attributed for the most part to maternal and sibling cannibalism. This would seem to be an example of parental care going wrong, although the cannibalistic habit may only be adopted when other food sources are scarce.

Parental care is shown, par excellence, by the ants and termites, by solifugids, by spiders, and, as already noted, by *Hemilepistus reaumuri*. It also occurs, in a rather attenuated form, in *Sphincterochila zonata*. This snail deposits clutches of eggs in the soil, and each clutch is laid in a mucus bag that is then sealed off (Shachak, Orr, and Steinberger 1975). This type of behavior is reminiscent of some New World solifugids that lay eggs in soil burrows that are subsequently sealed. Clearly, this is a device for reducing egg mortality due to desiccation. In addition *S. zonata* reduces the risk of complete reproductive failure by being iteroparous, and, indeed, this mollusk has been observed to deposit 17 egg clutches within a single reproductive season (Shachak, Orr, and Steinberger 1975). The sympatric *Trochoidea seetzeni* may also produce more than one clutch per season (Yom-Tov 1972), but neither in this species nor in *S. zonata* is there any evidence that iteroparity is associated with small clutch size. The number of eggs in a clutch is very variable, from 4 to 62 in *S. zonata*, and assuming an average of 30, this level of fecundity is on a par with many iteroparous mesic mollusks (Peter Newell 1980: personal communication). On the other hand, Yossi Steinberger (1980: personal communication) has found that *Sphincterochila prophetarum* produces relatively few eggs at any one time.

Egg survival is also very variable from clutch to clutch. Field observations by Shachak, Orr, and Steinberger (1975) on *S. zonata* revealed that egg survival can vary from less than 1 percent to nearly 70 percent, depending on the time of year when oviposition occurred and, by inference, the climatic conditions. Evidently, this mollusk is unable to always select the most suitable sites for egg laying, but repeated deposition of clutches in a range of sites over a single season increases the prospect of some egg survival. Aspect features prominently in the reproductive success of both *S. zonata* and *T. seetzeni*. Populations of these two species living on south-facing slopes in the Negev produced significantly larger clutch sizes than did populations on north-facing slopes (Table 9.2). Experiments carried out on *T. seetzeni* by Yom-Tov (1972) showed that a density-dependent control of fecundity occurs but operates at different levels on the two slopes. The percentage of snails laying eggs increases with density up to a limit of 2.8 individuals m^{-2} and thereafter declines on north-facing slopes. This limit is reached at average densities of 1.2 individuals m^{-2} on southern slopes. Furthermore, in the latter case but not the former, clutch sizes increased at lower densities. Clearly, these two populations, which are morphologically identical, behave differently as far as their reproductive tactic is concerned, and Yom-Tov (1972) provides some evidence to suggest that this difference is genetically based. Certainly, the two populations are subjected to different selection pressures, for mortality among young snails is much higher on the warm, dry south-facing slopes than on those facing north. This mortality is partly attributable to climatic factors but also in a considerable measure to predation by the dormouse *Eliomys melanurus*, which is more intense on the well-vegetated southern slopes than on north-facing ones. In view of this fact the reproductive tactic of *T. seetzeni* makes sense in that it responds to low population densities by increasing the proportion of the population laying eggs and by increasing the clutch size. *Sphincterochila zonata* evidently behaves in a similar manner (Yom-Tov 1970). However, according to Yossi Steinberger (1980: personal communication), predation by rodents on *S. prophetarum* would seem to be minimal.

TABLE 9.2.
Densities of Eggs and Adults of *Trochoidea seetzeni*
in an Area of 100 m^2 on North- and South-Facing Slopes of a Wadi

Slope Facing	Number of Adult Snails	Total Number of Eggs	Number of Eggs/Adults
South	108	1,794	16.6
North	495	964	1.95

Source: Data from Yom-Tov 1972.

These studies on desert snails are instructive in that they teach us not to look for generalized patterns of reproductive behavior but rather to appreciate the subtle differences that can occur within a particular population. The same may be true for desert isopods, and here, in the case of *Hemilepistus* species, parental care extends to guarding the young and sealing the entrance to the burrow with the body of the parent. This behavior not only prevents intruders from entering the burrow but also contributes to the maintenance of a high humidity in the burrow. Protected in this way, the young could be expected to show high survival rates; logically, *Hemilepistus* could be expected to have smaller clutch sizes than nonsocial desert isopods, such as *Armadillo officinalis*. However, it does not, as we have seen, although it is not immediately apparent why this should be so. Once again, it may be a reflection of the inability of this desert soil animal always to select a site that will allow the burrow to be excavated to the required depth. The production of a large clutch size will compensate for the extinction of families that have been unsuccessful in this regard.

PHYSIOLOGICAL ADAPTATIONS

So much emphasis has been placed, in this book, on what might be called the ecological physiology of desert soil animals that to return to this topic as a finale would seem to be laboring it. However, a synthesis should present an overview in which this category of adaptation can be considered alongside morphological and behavioral adaptations in the overall content of survival. Indeed, survival depends on a combination of all three of these types of adaptations, and in ecological terms survival is the ultimate criterion of success. Accordingly, we can now examine the extent to which soil animals living in arid environments show physiological convergence in response to the exigencies of their environment.

In this environment, as we have noted, rainfall is the driving variable for biological activity, but this activity is finely tuned by the temperature variable. Water and temperature are, then, the limiting factors, and physiological adaptations must be designed to control levels of body water and body temperature.

Control of Body Water

Essentially, this control is exerted by balancing water loss from the body by water intake. This balance has to be achieved in a matter of days or, at the most, weeks in those groups that have only a short life-span, such as rodents, certain tenebrionid beetles, and woodlice. Longer-lived taxa, such as snails and millipedes, may experience longer periods of water deprivation but may rehydrate when environmental water becomes more plentiful. It is against the background of this varying time dimension that the associated physiological tactics must be viewed. But there is also a spatial dimension. And if we examine, first, the ways in which desert soil animals acquire water from their environment, one fact demands immediate attention. This is that environmental water exists in discrete compartments that are sharply delimited both spatially and temporally.

The Time Factor. Rainfall events may be predictable (to within a few weeks) or unpredictable (to within several years), depending on whether the desert is seasonal or aseasonal. In either event the rainfall arrives in packages quite strictly delimited from each other by intervening periods of drought. To survive in this kind of environment, an animal could be expected to show physiological characteristics that are, for the most part, opportunistic. It must be metabolically alert to take advantage of periods when feeding and reproduction can occur under the most favorable conditions. The examples that immediately spring to mind, are *Orthoporus ornatus*, *Sphincterochila zonata*, and *Scaphiopus couchii*.

Quite apart from seasonal rainfall, environmental moisture arrives in the desert soil system in packages that are separated by much shorter intervals of time, notably through dew formation, which occurs during the 24-hour cycle for considerable periods of the year. This may provide an important source of moisture, as yet underestimated, for many desert soil animals. Certainly, *Hemilepistus reaumuri* takes advantage of this moisture source (Shachak, Chapman, and Steinberger 1976), as do the North African and Middle Eastern snail *Eremina ehrenbergi* (Billingham 1963), the sand roach *Heterogamia* (Ghabbour, Mikhail, and Rizk 1977), and possibly cryptostigmatid mites of the subterranean fauna. The intake of condensed moisture with the food items of granivorous ants and rodents may also be considered an exploitation of this moisture source. The ability to cash in on this supply of environmental moisture may allow extended periods of activity beyond the immediate limits of the rainfall season, particularly for those animals that can seek refuge within the soil during the hottest parts of the day. The metabolic implications of this may be quite different from those obtaining, for example, in the case of *Orthoporus ornatus* and *S. zonata*, which are placed in a situation of greater environmental stress over a longer period of time.

The Spatial Dimension. Rainwater and dew represent environmentally free water, which is available to animals that can drink, either orally or anally. Under saturated conditions free water may also be present in the atmosphere, and animals that can absorb this moisture across their body surface can avail themselves of this water source. These three types of free water constitute a spatial compartment—distinct from others, as we will see—in desert ecosystems that is exploited in different ways by different groups of the soil fauna. The following examples illustrate this point. Scorpions, centipedes, the vinegaroon *Mastigoproctus giganteus*, and the millipede *Orthoporus ornatus* have the common ability to imbibe free water by oral uptake; this is the only way in which scorpions and centipedes can exploit this particular source of moisture. However, *M. giganteus* evidently can also absorb atmospheric moisture across its body surface in saturated conditions (Ahearn 1970b), and this ability it shares with *O. ornatus* (Crawford 1978). Finally, *O. ornatus* can also tap free water by anal drinking; with this capability it must be considered as more versatile than these other taxa.

Of course, scorpions, centipedes, and uropygids, like all carnivores, obtain a considerable proportion of their water requirements from the body fluids of their prey. This particular compartment, preformed water, is broadly linked with the previous one since much of this water originally entered the desert system as free water. This also applies to the water in plant tissues consumed by herbivores. Plant water is present in considerable quantities in cacti and succulents, but the high concentrations of salts in desert plants present problems. Animals that rely on this water source have to have physiological adaptations to cope with the excessive amounts of salts that are likely to be absorbed with the plant water. Such animals include the pack rat *Neotoma*, which feeds on cacti; the sand rat *Psammomys*, which grazes halophytes; and, probably, also the honey-pot ants, which utilize the tissue water of cacti. These cactus feeders have to metabolize the oxalic acid present in their food and eliminate large amounts of calcium (Schmidt-Nielsen 1964).

A third compartment is provided by chemically bound water in the food, which can be released during metabolism. Granivorous rodents certainly exploit this particular package; kangaroo rats satisfy virtually all their moisture requirements from this source. *Orthoporus ornatus* is probably also able to obtain water in this way, although it is unlikely that it could generate enough for its overall needs on an annual basis. Metabolic water supplements that from other sources, in the case of this millipede, and this may be a widespread phenomenon among the desert soil fauna in general.

Finally, water in the form of atmospheric vapor is available to those animals that can absorb it under subsaturated conditions. This ability would seem to be rare among desert soil animals, and the two species in which it has been documented, the roach *Arenivaga investigata* and the thysanuran *Ctenolepisma terebrans*, clearly have physiological adaptations that have evolved independently. *A. investigata* employs aborptive surfaces derived from the foregut (pharyngeal bulbs), whereas *C. terebrans* absorbs atmospheric water vapor through the wall of the rectum.

From this kind of review, brief though it is, emerges the idea that environmental water exists in discrete packages, separated but not entirely unrelated in time and space, and that these packages are exploited to different extents and by different adaptations by the desert soil fauna. Some taxa, and possibly *Arenivaga* is an example, may be able to avail themselves of water from all four of the compartments defined above. Others may rely heavily on only one source, and *Dipodomys* would seem to illustrate this. Many other members of the desert soil fauna are probably located between these two extremes. The position of any given group on this spectrum of physiological adaptations might provide an indication of how successful it is as a desert animal.

On the other hand, the criterion for success could lie in the ability to conserve body water. Some water loss is inevitable in an arid environment, and this has attendant problems, such as the increased viscosity of body fluids and salt imbalance. In examining these problems further, it is perhaps useful to recognize that, just as with environmental water, body water occurs in different compartments and is lost from these compartments in different ways, at different rates, and under different conditions, depending on the animals in question.

Initially, a distinction can be made between water that occurs in the blood and that which is packaged in the body tissues. This distinction, however, is too simplistic for our purposes since the fate of tissue water varies with the location of the tissue within the body. The water contained in cutaneous and subcutaneous tissues may be lost from the body principally through transpiration. Water present in tracheae or lung tissue is lost through respiratory activities, notably during ventilation. The gut and its contents also contain water that can be lost via the feces. Much of this water is intercellular, and although this can be distinguished from intracellular water at the physiological level, a good deal of transfer may occur between these two compartments. Hemolymph or plasma water represents a third major physiological compartment that provides water for other compartments as this becomes depleted. Blood water is also lost from the body with the excretion of metabolic wastes, and it may also be incorporated into various secretions, such as repugnatorial substances, poisons, and pheromones.

Water that is lost from the general body surface is generally considered as transpiration water. As far as invertebrates are concerned, it is not usually practicable to separate out from this the respiratory water lost through the spiracles. The control of water loss through this particular pathway is exerted by spiracular-closing mechanisms in many tracheate arthropods. Additional regulation is achieved in the tenebrionid beetles by the presence of a subelytral space, which also serves, in a more general way, to reduce the vapor pressure gradient across the body surface. These features, which are more properly considered as structural adaptations, combine with others, such as a relatively thick integument that may be coated in lipid (nematodes, insects, millipedes, arachnids, and possibly the woodlouse *Venezillo arizonicus*, for example) or may not (many woodlice and centipedes). The absence of sweat glands in rodents and the thickened shell of desert snails can also be considered to belong to this same category of adaptations. All of these desert animals have in common a low basic rate of transpiration when compared with their mesic relatives. They also have physiological adaptations, as we have seen, to reduce water losses from the gut (by producing dry feces) and excretory systems (by producing a concentrated urine). These adaptations will act in concert to minimize the drain on hemolymph (or plasma) water. In this connection it is worth recalling that desert toads can transfer water from the bladder compartment to the blood and tissues. However, some loss of water from the blood may be inevitable, particularly during long periods of exposure to drought conditions when water in the cuticle/tissue compartment may be depleted. Under these conditions other physiological back-up systems may come into operation. These are designed either to allow the animal to tolerate variations in hemolymph concentrations or to regulate the concentration of osmotically active substances in the hemolymph as desiccation proceeds. The former tactic is employed by desert woodlice (Edney 1968a), *Scolopendra polymorpha* (Crawford, Riddle, and Pugach 1975), the scorpions *Paruroctonus aquilonalis* and *Hadrurus arizonensis* (Hadley 1974; Riddle, Crawford, and Zeitone 1976), and the testudine *Gopherus agassizii* (Dantzler and Schmidt-Nielsen 1966). Osmoregulation, on the other hand, appears to have been adopted by the roach *Arenivaga* (Edney 1968a), the tenebrionid *Eleodes hispilabris*, and the millipede *Orthoporus ornatus* (Moffett 1975; Pugach and Crawford 1978). However, it must be pointed out that *O. ornatus* can tolerate variations in hemolymph osmolality, depending on sex and duration of desiccation (Riddle, Crawford, and Zeitone 1976). This limited range of examples indicates two alternative solutions to the problem of blood water loss; this is an area that deserves further attention in relation to other groups of desert soil animals.

Control of Body Temperature

Soil animals in hot and cool deserts run the risk of exposure to temperatures at the upper and lower limits of their tolerance range. They may have to cope with these in some way if they are to survive. There are also spatial and temporal patterns of environmental temperature conditions that correspond in some respects to the compartments of environmental moisture discussed above. However, there are certain differences of emphasis, and these can now be briefly elucidated.

Patterns in Time. Patterns of temperature in the temporal dimension are expressed both in the long term (seasonal) and in the short term (circadian). Theoretically, at least, in both of these time scales biological activity should be bimodal since the distribution of the most favorable temperature conditions is bimodal. On a seasonal basis these conditions occur during the spring and autumn in hot deserts, with intervening periods of low (winter) and high (summer) temperatures. On the diurnal scale the most favorable temperatures will occur in the early morning and the late afternoon, and these are separated by intervals of low (nightly) and high (daily) temperatures.

Some groups of desert soil animals evidently show bimodal peaks of activity during the annual cycle. These include the heteromyid rodents, which estivate during the summer and hibernate during the winter but are active at other times; the same may well be true for many desert soil insects, such as tenebrionid beetles, ants, and termites. Temperature may have an important role in regulating the activities of these groups on a month-to-month basis, even though their overall response to desert conditions may be triggered by unimodal rainfall events. This rainfall trigger would seem to be the overriding factor in determining the response of the millipede *Orthoporus ornatus* and the snail *Sphincterochila zonata*. On the other hand, circadian activity patterns may be directly related to the temperature factor. The cessation of surface activity during the forenoon and its resumption during the late afternoon by ants, millipedes, tenebrionid beetles, lagomorphs, and certain rodents illustrates this fact very well. As noted earlier, these patterns may shift with the seasons, extending into the day or night as the period of favorable temperature lengthens with the season. Even so, the activity patterns of many groups of desert soil animals are sufficiently fixed to allow them to be classified as nocturnal, diurnal, or crepuscular.

Spatial Patterns. These patterns of temperature have both horizontal and vertical components. The former are expressed by variations in patterns of sun and shade, and these achieve a special significance when it is realized that soil surface temperatures are lower than ambient in shaded areas, whereas they are higher than ambient in exposed sites. Again, vertical temperature gradients are perhaps more strongly developed than moisture gradients, and the comparatively equable temperature conditions occurring below the soil surface provide an escape from the widely fluctuating surface temperatures.

Desert soil animals, in general, exploit these horizontal and vertical patterns by positioning themselves in such a way as to minimize temperature stress. The behavioral adaptations that allow them to do this—burrowing, nocturnalism, and orientation—have already been considered earlier in this chapter, but it is perhaps worth emphasizing that behavioral thermoregulation plays an important role in the survival strategies of desert animals. We have also reviewed some of the structural features that will combine with these behavioral adaptations to reduce the radiation load from the environment. It is now time to turn our attention to the physiological adaptations that are also involved. In doing this it is worth recalling that desert soil animals may have to adapt not only to high temperatures but also to low ones. These two topics are considered separately.

Adaptations to High Temperatures. These can be divided broadly into two categories: those that permit a tolerance to thermal loading and those that regulate this loading.

It is not surprising that the poikilotherms among the desert soil fauna (invertebrates, amphibians, and reptiles) can tolerate fluctuations in body temperature. Generally, these animals have higher lethal temperatures than their mesic relatives (see Chapter 7), and the ability to carry the additional heat load that this implies obviates the necessity for energetically expensive regulation. What is perhaps surprising is that desert homiotherms (rodents and marsupials) show similar types of tolerance (see Chapter 7). Here we have a hint of a common tactic adopted in physiologically quite dissimilar taxa.

Some groups of desert soil invertebrates exhibit physiological thermoregulation by metabolic compensation at high temperatures. This is indicated by relatively low Q_{10} values between 20°C and 40°C recorded for woodlice, the millipede *Orthoporus ornatus*, the uropygid *Mastigoproctus giganteus*, the scorpions *Hadrurus arizonensis* and *Paruroctonus utahensis*, and the spider *Lycosa carolinensis* (Edney 1964b; Ahearn 1970b; Hadley 1970b; Moeur and Eriksen 1972; Wooten and Crawford 1974; Riddle 1978). The mechanism behind this compensation, suggested by Wooten and Crawford's studies on *O. ornatus*, involves a switch from fat to carbohydrate as the metabolic substrate during the hot summer period (see Chapter 7). However, the universality of metabolic compensation as a thermoregulatory device in desert soil animals has been called into question by Clifford Crawford (1980: personal communication). Be this as it may, this solution to the problem of overheating would seem to be more realistic, as far as animals living in arid environments are concerned, than thermoregulation through evaporative cooling, which would deplete body water levels (see earlier).

Adaptations to Low Temperatures. Soil animals in hot deserts seem less capable of dealing with the problem of cold than with that of heat. Studies in the cold polar desert of Antarctica (Block 1980) suggest that soil invertebrates have adopted one of two possible tactics: they are either freeze-tolerant or freeze-susceptible. Immediately, we can inquire as to whether the soil animals of hot deserts can be assigned to one or other of these two categories.

The work of Zachariassen and Hammel (1976) has shown that

hot desert tenebrionid beetles of the genera *Coelocnemis* and *Eleodes* that inhabit montane environments in southern California are represented by freeze-tolerant and freeze-susceptible forms. The same can be said for desert scorpions: *Diplocentrus spitzeri* is freeze-tolerant, whereas *Paruroctonus utahensis* from desert grassland is freeze-susceptible (Crawford and Riddle 1974; Riddle and Pugach 1976). As Block (1980) has shown, freeze-susceptible arthropods in the Antarctic are capable of supercooling to a much greater extend than freeze-tolerant ones, and the same appears to hold true in hot deserts. *P. utahensis* shows a seasonal pattern of supercooling to a point around $-12°C$ in autumn (Riddle and Pugach 1976), which should ensure adequate survival through the winter in the southwestern deserts of the United States. In these same deserts *D. spitzeri* shows no such seasonality (Crawford and Riddle 1975), and although this scorpion is moderately frost tolerant, its habit of digging only a shallow burrow combines with its inability to supercool below about $-7°C$ to increase the risk of winter mortality (see Chapter 7). The same is evidently true for the desert centipede *Scolopendra polymorpha* (Crawford, Riddle, and Pugach 1975).

Block (1977, 1980) also reported that Antarctic microarthropods have elevated metabolic rates at temperatures in the range of $-4°C$ to $+15°C$. This tactic for cold adaptation has either not been discovered or does not exist in the invertebrates of hot deserts. Indeed, the overwintering physiology of the vast majority of hot desert soil animals requires much more attention before any patterns, if such there be, become apparent.

10
Life-Styles for Survival: Strategies

It is fashionable to identify suites of adaptations (tactics) with an overall strategy that can be located within the continuum of "r" and "K" selection (MacArthur and Wilson 1967; Pianka 1970). The theoretical basis and, indeed, the practical applicability of the r and K continuum concept have been evaluated and reevaluated, analyzed and criticized frequently enough (see, for example, Southwood 1976; Stearns 1976, 1977; Block 1980), and no good purpose would be served by tracking down a pathway of theory already well worn. Nor should it be necessary to belabor the various attributes of r and K strategists since these are adequately documented in the literature. However, it does seem appropriate (since this has not so far been attempted on a general basis) to examine the life-styles of the desert soil fauna in the context of the r and K continuum. To do this, some basic parameters must be identified; these have been enumerated in a most refreshing manner by Horn (1978), and they are illustrated in Figs. 10.1 and 10.2. With this author we can readily appreciate that the two diagrams are not mirror images of each other, and it follows, therefore, that r selection and K selection are not just opposite ends of a one-dimensional spectrum of strategies. Such strategies may implicate a second dimension that P.J.M. Greenslade (in preparation) has termed "adversity selection." We will return to this point later, although it is pertinent to remark on it here since a slavish adherence to the idea of a unidimensional spectrum of r and K strategies has brought its share of detractors. The concept of adversity selection may provide at least a partial answer to these critics.

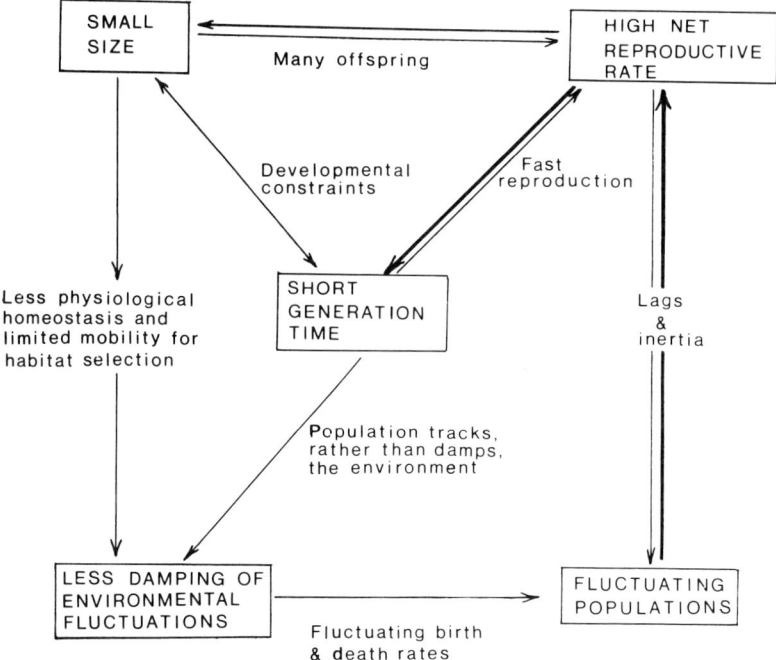

Fig. 10.1. Positive feedback loops associated with extreme r selection. Heavy arrows represent actual selection. (From Horn 1978.)

A perusal of the schemata presented in Figs. 10.1 and 10.2 will identify a number of parameters that merit attention:

1. Body size,
2. Environmental predictability,
3. Response to the environment,
4. Reproductive rate,
5. Generation time and longevity, and
6. Population changes.

Each of these will now be considered in turn.

BODY SIZE

It has been argued that animals of relatively large body size should be K strategists, and those of relatively small body size should be r strategists. The inference that has been drawn from

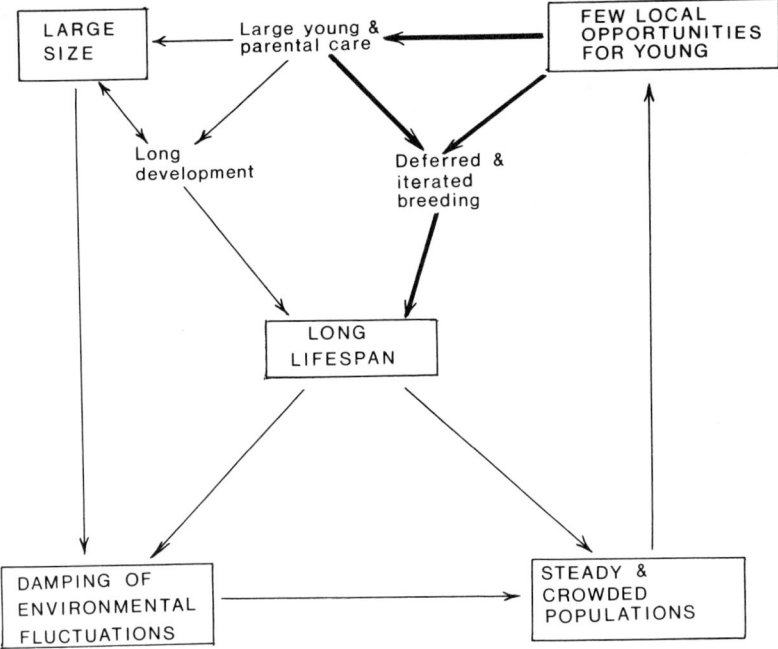

Fig. 10.2. Positive feedback loops associated with extreme K selection. Heavy arrows indicate actual selection. (From Horn 1978.)

this is that vertebrates are mainly K selected, whereas invertebrates are r relected. This tenet seems so much of an oversimplification as to have little ecological meaning. It is perhaps more perspicuous to make comparisons of body size within, rather than between, taxa, and this can be done for a number of groups of desert soil fauna. The advantages of large body size have already been considered (Chapter 9), and there is no doubt that desert tenebrionid beetles, millipedes, and centipedes are larger than their mesic relatives. Similarly, the desert scorpions *Paruroctonus mesaensis* and *Hadrurus arizonensis* are among the largest of the North American scorpions, and the former, at least, has many attributes of a K strategist (Polis and Farley 1980). However, there is no consistent trend toward selection for large body size in many other taxa (woodlice, snails, ants, and microarthropods, for example), and we must conclude that this is no absolute determinant of a particular survival strategy. It may be important for certain taxa that exhibit features associated with K selection, but it does not appear to figure prominently in the life-styles of others that, for other reasons, may be K or r selected.

ENVIRONMENTAL PREDICTABILITY

The unpredictability of the desert environment is so well recognized as to be almost axiomatic. Intuitively, it is to be expected that there will be a strong selection in favor of r strategists. However, this expectation ignores the possibility that the immediate environment of a desert soil animal may be a good deal more stable than that of the region as a whole. This is particularly true of those animals that construct permanent burrows and nests, often deep in the soil. The humidity and temperature within these systems is usually subject to only minor diurnal and seasonal fluctuations, within the tolerance limits of the animals concerned. To illustrate this, we can return to the sand scorpion *Paruroctonus mesaensis*, which has been studied intensively in the dunes near Palm Springs, California, by Polis (1979, 1980) and Polis and Farley (1979, 1980). This predatory arachnid spends much of its life in a subterranean burrow; it is active on the soil surface for a period of only 4 to 8 hours during any 24-hour cycle, and then only at night. This activity pattern ensures that the species never ventures into an environment that has a large unpredictable component. To be sure, it is exposed to random predation during its surface-active periods, but this is minimized by reducing the duration of its forgaging activities. Its abiotic environment, on the other hand, is much less stochastic, and this supports the view that *P. mesaensis* is K selected. Other groups of desert scorpions that can be identified as K strategists by virtue of their deep-burrowing habit are *Buthus occitanus*, *Leiurus quinquestriatus*, and *Scorpio maurus*. In contrast the shallow-burrowing *Diplocentrus spitzeri* and the crevice-dwelling *Centruroides sculpturatus* are likely to experience more unpredictable environmental conditions; in these cases we can look for elements of an r-selected strategy.

Well-defined subterranean burrows and nesting sites are also a feature of the survival strategies of ants, termites, millipedes, the isopod *Hemilepistus*, and rodents. The environmental stability that accrues from such burrowing behavior suggests that the animals concerned are K selected. This is probably true of the majority of species that construct well-defined refuges in the soil, although woodlice of the genera *Hemilepistus* would seem to be an exception in this respect. As we have seen, the selection of a suitable burrow site is a risky business as far as *Hemilepistus* is concerned, and the unpredictability of success may result in r selection. Not so with the rodents studied by MacMillen and Christopher (1975), however. These authors found that the heteromyid *Dipodomys merriami*, *Perognathus longimembris*, and *P. fallax* and the cricetid *Neotoma lepida* and *Onychomys torridus* remained in positive states of water balance even during a period of severe drought. This indicates that these rodents rarely experience water stress, and this is consonant with K selection.

The risk factor increases in importance with the amount of time spent on the soil surface, particularly during the daylight hours. Animals that exhibit this type of activity should show characteristics of r selection and might be expected to include snails, ground squirrels, certain tenebrionids, and millipedes. However, the last two groups have already been considered to be K strategists in respect of their large body size. Both groups can produce repugnatorial secretions to deter predators, and the millipede *Orthoporus ornatus*, at least, only ventures onto the soil surface during a very limited and climatically most favorable period of the year. In these ways the unpredictability of the immediate environment may be minimized. The same effect may be produced by a strictly nocturnal activity pattern adopted by many groups that live in rock crevices or shallow, ill-defined burrows, such as many of the woodlice, sand roaches, some tenebrionids, and the scolopendromorph centipedes.

RESPONSE TO THE ENVIRONMENT

The impact of environmental fluctuations can be reduced by physiological mechanisms designed to maintain homeostasis and by a mobility that enables an area to be reconnoitered for suitable habitats and food.

Physiological homeostasis is largely mediated by the curtailment of body water loss, and it will have become apparent, from the discussion presented in Chapter 6, that virtually all groups of desert soil animals, for which data are available, show lower rates of water loss than their mesic relatives. Even when empirical data are lacking—as, for example, in the soil-dwelling mites and collembolans—the strong representation of genera known to be xerophilic points to selection for physiologically adapted groups. However, this knowledge does not advance the argument any since all it shows is that desert animals are better adapted to life in deserts than are mesic animals, and this must be self-evident. More worthy of attention is the finding that different groups of the desert soil fauna have solved the problems of physiological homeostasis with differing degrees of success.

In deploying the argument further it is convenient to turn, once again, to the large predatory arachnids of hot deserts. Rates of body water loss in various groups of these arachnids are given in Table 6.1, and they indicate three levels of physiological response that can be interpreted as high, medium, or low. The high response is exemplified by *Hadrurus arizonensis, Leiurus quinquestriatus,* and *Androctonus australis;* the medium response is exhibited by *Centruroides sculpturatus* and the solifugid *Galeodes arabs;* the scorpion *Diplocentrus spitzeri,* the uropygid *Mastigoproctus giganteus,* and the araneid *Eurypelma* rank lowest in their ability to restrict body water loss. It is tempting to suggest that the high-response group can be identified as K strategists, the low-response group as r strategists, with the remainder somewhere in between, although this may be too simplistic. We have seen, though, that *H. arizonensis* and *L. quinquestriatus* have other attributes of K strategists, while *C. sculpturatus* may be r selected, so that a pattern may be developing. Moreover, *D. spitzeri* appears to have little physiological defence against cold (Chapter 7), which will render this species liable to large seasonal fluctuations in population size. The same may be true of the centipede *Scolopendra polymorpha.* Unlike the scorpions this myriapod lacks a layer of waterproofing lipid in its cuticle, which further adds to the difficulties of achieving homeostasis.

Desert woodlice also do not have a waterproofed cuticle, as a general rule, although *Venezillo arizonicus* may be exceptional in this respect. *V. arizonicus* is considered to be rare, and it may be that this species occurs in low but stable densities in the deserts of the southwestern United States. In contrast *Hemilepistus reaumuri* has little physiological defense against drought and, once again, is seen as a product of r selection. *Armadillo* spp. share with *Venezillo* the ability to conglobate; members of both these genera lack tegumental glands, and these features will allow effective control of body water loss, as befits a K strategist.

Water loss is also controlled efficiently by such snails as *Sphincterochila zonata* and *Otala lactea* (see Chapter 6), and this can be interpreted as a positive response to the dangers of water stress.

The millipedes and insects among the desert soil fauna would seem to be well adapted to withstand physiological stress by virtue of their waxy cuticle, ventilation control mechanisms, a reflective body surface, and, in *Orthoporus* and *Arenivaga* at least, the ability to regulate the osmotic pressure of the hemolymph. Here again are elements of K selection, although the termites are perhaps a special case since they are prone to desiccation. However, these social insects are very largely subterranean in habit in hot deserts and are rarely exposed to physiological stress.

Special cases are also provided by nematodes, Collembola, and possibly prostigmatid mites, which increase their resistance to physiological stress by assuming some form of anhydrobiosis. In a sense this is the very opposite to homeostasis, but its effect is very much the same. Also included among this category of adaptations may be the ability, shown by desert marsupials, for example, to tolerate variations in body temperature. All of these features increase the probability of survival in stressful situations and reduce the possibility of large fluctuations in population densities.

Mobility is very much associated with foraging activity; scorpions, ants, rodents, sand roaches, and tenebrionid beetles illustrate this very well. All of these groups are highly mobile within the soil as well as on its surface, and selection for K strategy seems to be present in at least some members of each group. Enhanced mobility increases foraging efficiency, which ensures that the minimum amount of time is spent on the soil surface where an animal may be vulnerable to predators. This tactic of time minimizing is a feature of K selection and is well illustrated by the scorpion *Paruroctonus mesaensis* (Polis and Farley 1980). Of course, highly mobile and highly efficient foragers are likely to encounter competitive situations, and the mechanisms that operate to ensure resource allocation in these situations have already been considered in Chapter 8. Competition is keen among (and between) granivorous rodents and ants in hot deserts (Chapter 8) and also between scorpions where it may operate to regulate population sizes in a density-dependent manner (Polis and Farley 1980). Such regulation is consistent with K selection. Competition is also intense among the more completely subterranean rodents (Chapter 4), which occur in low but stable densities.

Some groups show only a limited mobility, however, and here we can include the snails and woodlice. As a consequence they are subject to heavy predation: *Hemilepistus* by the scorpion *Scorpio maurus* in the Negev (Shachak 1980), snails by rodents and birds (Yom-Yov, 1972; Yossi Steinberger 1980: personal communication). This factor may contribute to erratic fluctuations in population densities, particularly in suboptimal habitats.

REPRODUCTIVE RATE AND LONGEVITY

In theory there are two alternative reproductive tactics open to desert soil animals that occupy opposite ends of the r/K spectrum. First, there is the boom-or-bust phenomenon, characteristic of r strategists, in which a large number of young are produced during a particularly favorable period. Many of these young will die before reaching maturity, either through predation or the inability to defend themselves against the extremes of their physical environment. Some, however, by the laws of chance will survive to perpetuate the species. This tactic places a premium on production, and a considerable amount of energy is invested in reproduction, with little left for adult growth. Consequently, the adult stage will tend to be short-lived and often produces only one brood (semelparity). A prime tactician in this context is *Hemilepistus reaumuri*, and it is also likely that scolopendromorph centipedes, solifugids, uropygids, and cryptostigmatid mites belong in this category (see Chapters 3 and 4). The alternative tactic is to channel much of the available energy into long and sustained growth, not only to maturity but also once the adult stage has been reached. As a consequence relatively few young are produced at any one breeding period, but time may be available for successive broods to be produced (iteroparity). These features, characteristic of K strategists, are found among certain rodents, scorpions, tenebrionid beetles, and social insects. However, there are enough examples among the desert soil fauna that do not fit neatly into one or other of these two categories of selection to suggest that the distinction between them is too simply drawn and may not have a generality that is ecologically meaningful. The scorpion *Urodacus yaschenkoi*, for example, appears to invest so much energy into growth toward maturity that it has relatively little left to devote to adult growth or reproduction. Indeed, semelparity in this species is associated, unexpectedly, with a low clutch size, and it must be inferred from this that mortality is low among juveniles.

In an unpredictable environment, such as a hot desert, juvenile and adult mortality will be subject to a good deal of variability, and it may be important to distinguish between these two mortality categories in attempting to define strategies. If adult mortality is variable, there is no certainty that this stage will survive to breed more than once, and selection will favor early breeding and a large clutch size. In other words the species adopting this tactic are bet-hedging in the sense that the production of a large number of young at any one time has a large, built-in margin for failure; many will die, but some will survive. On the other hand, if juvenile mortality is variable, a species can hedge its bets by a more controlled reproductive response. Long-lived adults can breed more than once, and the production of several to many egg batches, often deposited in a variety of microhabitats, spatial and temporal, spreads the mortality risk and ensures the survival of some.

In practice early breeding and large clutch sizes are not

inseparably linked. The woodlouse *Porcellio olivieri*, for example, breeds as early as possible in North Africa, but it does not produce a notably large clutch size. This semiarid species occurs where suitable habitats, in the form of organic debris, are not hard to find; these not only afford shelter for breeding adults but also sites for colonization by dispersing young. The prospect of enhanced survival afforded by these sites obviates the necessity for large brood sizes, even in this relatively short-lived species.

If semelparity is not necessarily associated with large clutch size—and clearly it is not in the case of *Urodacus yaschenkoi*—then perhaps we should question whether iteroparity is always associated with small clutch size. The short answer is that it is not. To be sure, there are instances where this relationship holds. For example, the snail *Sphincterochila prophetarum* has a long period of maturation (four to five years) and, in this respect, contrasts with *S. zonata* (Yossi Steinberger 1980: personal communication). It combines iteroparity with a low annual egg production, indicative of a K strategist. Predation pressure on this species by thrushes is relatively high during the active period, and it compensates for this by spreading its reproductive effort. However, the iteroparous *S. zonata* and *Trochoidea seetzeni* produce egg clutches that are neither remarkably large or small. Again, Clifford Crawford (1980: personal communication) considers egg production to be considerable in *Orthoporus ornatus* and *Archistreptus* sp. There appears to be selection against reducing clutch size, despite iteroparity, in these millipedes and snails. This may be a consequence of predation pressure on juvenile stages, which are surface-active in both of these groups; it may also reflect a relatively short life-span or maturation period, at least in the case of *S. zonata*.

These few examples highlight the difficulty of locating various reproductive tactics within a one-dimensional r/K spectrum. The difficulty is compounded when other attributes are brought into the picture. For example, it will be evident from what has been said earlier in this chapter that the scorpions *Buthus occitanus* and *Leiurus quinquestriatus* are K selected as far as their ability to minimize the impact of the physical environment is concerned. In addition *B. occitanus* has the K-selected feature of relatively large body size. However, neither this species nor *L. quinquestriatus* conform entirely to the ideal of a K strategist as far as their reproductive tactics are concerned. Both develop to maturity in a relatively short period of time and produce a larger-than-average clutch size (Polis and Farley 1979). These are features of r selection that suggest a bet-hedging tactic to counteract the dangers of a variable amount of adult mortality. Reliable data on this mortality in hot deserts are rather scarce, but perhaps it can be argued that mortality factors—and, therefore, selective pressure—may impinge more on surface-active adults than on subterranean juveniles in these deep-burrowing scorpions. However, this argument cannot be extended to the dune scorpion *Paruroctonus mesaensis*, which suffers a juvenile mortality of about 60 percent (Polis and Farley 1979). The response, in this case, is a rapid development to maturity, coupled with iteroparity and a relatively small clutch. Iterative breeding and the low energy costs associated with small clutch size are of selective value for this desert scorpion; perhaps the short period of postembryonic development may be a device for curtailing this, the most vulnerable interval in the life cycle.

The lesson to be learned from these scorpions is that generalizations about survival strategies cannot even be made within a particular taxonomic group, and this is further illustrated by the surface-active rodents. We may recall from Chapter 3 that these vertebrates, which intuitively could be regarded as K strategists, can be divided into three groupings on the basis of high, moderate, or low rates of survival. Heteromyids and sciurids are relatively long-lived and produce small broods. Murids, on the other hand, produce relatively large numbers of young that survive for only a short time; cricetids occupy a position intermediate between these two extremes. It seems logical to identify heteromyids, sciurids, and possibly cricetids as K strategists, while the murids may more properly be considered as r strategists.

POPULATION CHANGES

The culmination of the various factors discussed above—body size, action by and reaction to environmental features, reproductive tactics, and longevity—is expressed by the way in which populations change over a period of time. The boom-and-bust tactic of the r strategist can lead to population explosions during favorable conditions of food plenty and (relatively) benign climate. However, this is short-lived as the carrying capacity of the environment is rapidly exceeded, and population size decreases at least as quickly as it had previously increased. In contrast the more controlled reproductive tactic of the K strategist, coupled with its greater physiological resilience, ensures a more stable level of population density. This contrast, of course, reflects the investment of energy into reproduction by r strategists and into growth by K strategists. In closing this volume it may be useful to look briefly at some of the groups of desert soil fauna from this point of view. This survey is necessarily incomplete since demographic data are lacking for certain groups such as centipedes, sand roaches, tenebrionid beetles, psocopterans, pseudoscorpions, pauropods, enchytraeids, and insect larvae.

Scorpions

There is no evidence to suggest that population densities of these large-sized, mainly surface-active, predatory arachnids exhibit violent fluctuations. The most intensively studied species, *Paruroctonus mesaensis*, has many of the attributes associated with K selection, and although it may occur in high densities—up to 4,000 ha^{-1} according to Polis (1979)—density-dependent regulation occurs through cannibalism and predation, which stabilizes population levels. Such regulation suggests that this scorpion lives in a competitive situation, and this is probably true of many other desert scorpions.

Solifugids and Uropygids

Little is known of the population dynamics of these predators, although *Galeodes arabs* lives for only a year during which time it produces a large number of young (Cloudsley-Thompson 1961b). This is suggestive of an r strategist, although, like the scorpions, cannibalism may operate in a density-dependent fashion to reduce violent fluctuations in population size. More demographic data are needed concerning this group and the uropygids. The latter are never very common in deserts, and presumably they exist at low equilibrium levels.

Amphibians and Lizards

The mass emergence of anurans from the soil during rainfall events is not uncommon in hot deserts. It is tempting to equate this phenomenon with a population explosion, although in reality it represents reactivation from a dormant condition. Nevertheless, anurans have a high reproductive potential, and *Scaphiopus hammondi* may experience up to 50 percent mortality among overwintering juveniles in the Chihuahuan desert (Creusere and Whitford 1976; Whitford and Meltzer 1976). From these observations it is likely that population fluctuations may occur that are more appropriate to an r than a K strategist.

Lizards that are of particular interest in the context of this book are the horned lizards of the genus *Phrynosoma*, which prey heavily on ants. In the Chihuahuan desert *P. cornutum* feeds mainly on harvester ants of the genus *Pogonomyrmex* (Chapter 3), and data assembled by Whitford and Bryant (1979) indicate that this desert lizard is food limited rather than predator limited. It has an optimal foraging strategy that minimizes the risk associated with the procurement of food. To this extent, it is not an opportunist but rather an equilibrium species—in effect, a K strategist. The related *P. modestum*, which preys mainly on honey-pot ants of the genus *Myrmecocystus*, can invoke diet switching (Shaffer and Whitford 1981) to ameliorate potential food stress. These tactics are symptomatic of a conservative strategy that can be identified with K selection. To what extent this line of reasoning can be applied to desert lizards as a whole remains to be discovered.

Rodents

A considerable amount of demographic data is available for desert rodents, and the summary presented in Table 3.2 indicates that while heteromyids, sciurids, and, to a lesser extent, cricetids maintain equilibrium densities, and are, therefore, k strategists, the Australian murids do not. The population explosions that are such a feature of the life-styles of these murids have already been mentioned, and this boom-or-bust tactic is undoubtedly a response to the extremely harsh and unpredictable environment of the Australian arid zone. The murids, and probably also the soil-based marsupials, which have a high reproductive output, are evidently r strategists par excellence. Again, as noted in Chapter 8, some cricetids that occur only as transients in American deserts may show features of r selection.

Millipedes

The rather fragmentary data that are at hand concerning the population dynamics of desert and semidesert millipedes are difficult to fit into any general pattern; indeed, such a pattern may not exist. The Nigerian gomphiodesmid and paradoxosomatid millipedes evidently produce a large number of young during a restricted breeding season, and surface swarming occurs quite frequently. Such behavior is likely to result in the kind of population fluctuations characteristic of an r strategist with a short life-span. The much longer-lived desertic *Orthoporus ornatus* and *Archistreptus* sp. might be expected to show more of the features of K selection, and, as we have seen, this is true in many ways.

Woodlice

Here again, generalizations may not be appropriate. The biology of *Hemilepistus reaumuri* has been discussed at some length in this book, and it is evident that it is an r strategist in many respects. Data relating to population changes in this species are not always easy to interpret, but the production of a large number of young by this short-lived species can be seen as a bet-hedging tactic to offset high levels of juvenile mortality. The risks involved in selecting a suitable burrow site are considerable for *H. reaumuri*, and it is to be expected that population densities will show marked fluctuations from year to year, despite parental protection of the young and territorial behavior; these remain to be documented, however.

There is no reason to believe that other desert woodlice follow the example of *H. reaumuri*. *Venezillo arizonicus* combines a low reproductive output with low population density—features of K selection. Other species may occupy positions intermediate between these two extremes, as indicated in Fig. 10.3, although population data are hard to come by in these cases.

Snails

Both *Sphincterochila zonata* and *Trochoidea seetzeni* show density-dependent control of their reproductive response, as we have already noted, and this in itself will reduce the tendency for population densities to fluctuate violently. At high densities clutch sizes are reduced, as is the proportion of reproductively active females in the population. At low densities these procedures go into reverse, and this tactic can be seen not only as an extremely efficient way of utilizing the energy available—by directing into growth or reproduction as the occasion demands—but also as a way of counteracting the effects on population density of very variable juvenile mortality. Here is an example of a special kind of tactic of bet-hedging, identified by Stearns (1976), which is difficult to reconcile with either r or K selection. Stable population densities

258 Desert Soil Fauna

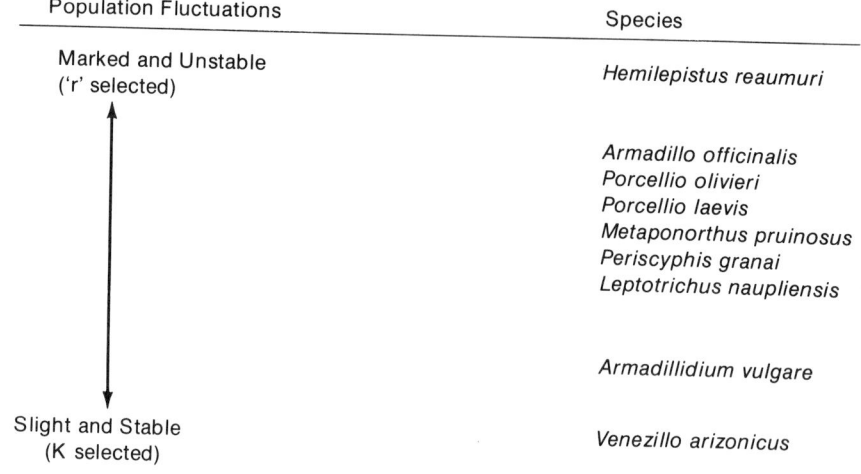

Fig. 10.3. Arid and semiarid wood lice arranged according to their possible population characteristics in an r/K continuum.

have also been observed in *Sphincterochila prophetarum* (Yossi Steinberger 1980: personal communication).

Insects

The social organization of the ants and termites ensures that colony growth proceeds in a controlled manner, but competitive encounters between colonies of the same and different species, particularly among ants, and predation are also likely to be intense in hot deserts. For example, the horned lizard *Phrynosoma cornutum* utilizes harvester ants of the genus *Pogonomyrmex* at or near the maximum exploitation level (Whitford and Bryant 1979). The related *Phrynosoma modestum* relies on honey-pot ants of the *Myrmecocystus depilis/mimicus* complex for nearly 90 percent of its food intake, although this predator shows diet-switching behavior (Shaffer and Whitford 1981). These depredations have a significant impact on ant numbers to the extent that Whitford and Bryant (1979) suggested that *Pogonomyrmex* spp. must have a yearly productivity almost equal to the standing crop biomass. Under these circumstances a large investment in reproduction is to be expected to counteract short-term losses. However, there is strong evidence that this predator/prey system is a product of coevolution, and, as such, population densities of both predator and prey are likely to show a measure of stability in the long term identifiable with K selection. Once again, we have recourse to only a very limited number of well-documented examples. Much more data are required concerning the productivity of desert ant species before these ideas can be tested further.

It could be expected that the above considerations would apply to desert termites, although, once again, it is difficult to glean much demographic data from the literature. There is evidence (see Chapter 3) that the densities and diversity of these insects are lower in deserts than in semiarid regions and savannahs. Nevertheless, mature colony sizes of between 5,000 and 10,000 individuals for the Sonoran desert termite *Gnathamiitermes perplexus* have been reported in the literature (Johnson and Whitford 1975). These estimates exceed, by rather larger margins, those of harvester ant colony densities given by Whitford and Bryant (1979). Using the data provided by these authors, it is possible to calculate that at peak periods *Pogonomyrmex desertorum* has a colony size of about 500; *P. rugosus*, 2,300; and *P. californicus*, 1,400.

In view of these elevated termite densities it is tempting to pursue the question of whether species packing is more prevalent among termites than ants in hot desert soils. This question has not been addressed, but it could be argued that the relatively low species diversity of desert termites allows for large niche breadth for those species capable of colonizing hot desert soils. If this is true, the stabilizing effects of competition should be less in evidence than in the case of the ants, and population fluctuations could be larger. Unfortunately, data to support or refute this suggestion are lacking.

Nematodes and Microarthropods

Our knowledge of the population dynamics of the surface-active component of the desert soil fauna is at best fragmentary. There is even less information available concerning the behavior of populations of the more completely subterranean taxa: nematodes, collembolans, mites, psocopterans, and pauropods.

Nematode densities in desert soils are appreciably lower than in mesic soils (Chapter 4). Although these soil animals may be capable of rapid turnover, their ability to assume an anhydrobiotic state suggests that they can survive in an unpredictable environment without resorting to a boom-or-bust tactic. Accordingly, it may be predicted that desert soil nematodes will not experience violent fluctuations in densities, at least on a short-term basis. There is here, then, a suggestion—but no more than that—that these animals are K strategists.

Wallwork (1980) has argued that cryptostigmatid mite populations in a Mojave desert soil show many of the features associated with r selection. Two species singled out for special attention—namely, *Joshuella striata* and *Haplochthonius variabilis*—have already been treated in Chapter 4, and it will have become evident that populations of both of these species respond by recruitment to the environmental triggers of rainfall and temperature. This suggests an opportunistic strategy, which is consonant with r selection. However, studies of this type with other groups of microarthropods are only just beginning.

CONCLUSION

What then, if anything, is to be made of all this? The data and observations presented in this and the previous chapter are capable of various interpretations, depending on which particular axe is being ground. There are, though, some points that emerge from this analysis that probably have a measure of validity; these can be stated rather simply.

In the first instance it is very clear that much more basic data need to be obtained relating to particular tactics and more general strategies adopted by virtually all groups of desert soil fauna. Intensive and, in some cases, extensive studies have so far been carried out on only a limited number of taxa: *Orthoporus*, scorpions, ants, and surface-active lizards and rodents, for example. Even here, there are important gaps in our knowledge. But if as much was known about desert woodlice, snails, nematodes, mites, collembolans, and pauropods, for instance, a clearer picture of the overall pattern of life-styles would become apparent. This is a difficult goal, but it is by no means impossible to attain. It is encouraging that research efforts by desert ecologists are being concentrated to redress this imbalance.

Second, the ideas embodied in the concept of r and K selection can be applied fairly confidently to certain groups of the desert soil fauna. In doing this it becomes obvious that selection in this unpredictable environment produces both r and K strategists. The former are represented by the wood louse *Hemilepistus reaumuri*, rodents of the family Muridae, and certain cryptostigmatid mites. Heteromyid rodents and scorpions are good examples of the latter. However, these generalizations cannot be applied to other groups. The tenebrionid beetles, for instance, probably include species that are r strategists and others that are K strategists. The same may be true of the woodlice, millipedes, and Collembola.

Finally, there are some groups that do not seem to fit easily into the r/K continuum. These probably include the snails and the termites. If these are indeed exceptions to the rule, they do not necessarily invalidate r and K theory. But they do point the way to the possibility that these two types of strategy are not the only ones open to animals living in the soils of hot deserts. Indeed, P.J.M. Greenslade (in preparation) has proposed an alternative that he terms "adversity," or A, selection. This would operate in harsh but predictable environments and would favor adaptation and the conservation of adaptation. It may well be that some of the examples we have been considering here accord with this type of selection since Greenslade postulates that desert animals have the ability to switch between r- and A-selected attributes.

References

Ahearn, G. A. 1970a. The control of water loss in desert tenebrionid beetles. *J. exp. Biol.* 53: 573-95.

——. 1970b. Water balance in the whip scorpion *Mastigoproctus giganteus* (Lucas) (Arachnida, Uropygi). *Comp. Biochem. Physiol.* 35: 339-53.

Ahearn, G. A., and Hadley, N. F. 1969. The effects of temperature and humidity on water loss in two desert tenebrionid beetles, *Eleodes armata* and *Cryptoglossa verrucosa*. *Comp. Biochem. Physiol.* 30: 739-49.

Anderson, J. M. 1975. The enigma of soil animal species diversity. *Proc. Vth. Int. Coll. Soil Zool.*, pp. 51-58.

Anderson, J. M., and Healey, I. N. 1972. Seasonal and inter-specific variation in major components of the gut contents of some woodland Collembola. *J. Anim. Ecol.* 41: 359-68.

Appleton, T. C.; Newell, P. F.; and Machin, J. 1979. Ionic gradients within mantle-collar epithelial cells of the land snail *Otala lactea*. *Cell Tiss. Res.* 199: 83-97.

Ayyad, M. A., and Ghabbour, S. I. 1977. Systems analysis of Mediterranean desert ecosystems of northern Egypt (SAMDENE). *Environmental Conservation* 4: 91-102.

Baker, G. H. 1978. The population dynamics of the millipede *Ommatoiulus moreletii* (Diplopoda: Iulidae). *J. Zool., Lond.* 186: 229-42.

Banerjee, B. 1967. Seasonal changes in the distribution of the millipede *Cylindroiulus punctatus* (Leach) in decaying logs and soil. *J. Anim. Ecol.* 36: 171-77.

Bar, Z. 1975. Distribution and habitat of the genus *Sphincterochila* in Israel and Sinai. *Argamon. Israel J. Malac.* 5: 1-19.

——. 1977. Range and habitat of the genus *Cristataria* Vest. *Argamon. Israel J. Malac.* 6: 1-16.

Bartholomew, G. A., and MacMillen, R. E. 1961. Oxygen consumption, estivation and hibernation in the kangaroo mouse *Microdipodops pallidus*, *Physiol Zoöl* 34: 177-83.

Billingham, J. 1963. Snail haemolymph: an aid to survival in the desert. *Aerospace Med.* 34: 643-47.

Block, W. 1977. Oxygen consumption of the terrestrial mite *Alaskozetes antarcticus* (Acari: Cryptostigmata). *J. exp. Biol.* 68: 69-87.

——. 1980. Survival strategies in polar terrestrial arthropods. *Biol. J. Linn. Soc.* 14: 29-38.

Blower, J. G. 1955. Millipedes and centipedes as soil animals. In *Soil zoology*, ed. D. K. McE. Kevan, pp. 138-51. London: Butterworths.

Bodine, M. C., and Ueckert, D. N. 1975. Effects of desert termites on herbage and litter in a short grass ecosystem in west Texas. *J. Range Management* 28: 353-58.

Bouillon, A. 1970. Termites of the Ethiopian Region. In *Biology of termites*, ed. K. Krishna and F. M. Weesner 2: 153-280. New York and London: Academic Press.

Brian, M. V., ed. 1978. *Production ecology of ants and termites*, International Biological Programme, vol. 13. Cambridge: Cambridge University Press.

Brody, A. R. 1970. Observations on the fine structure of the developing cuticle of a soil mite *Oppia coloradensis* (Acarina: Cryptostigmata). *Acarologia* 12: 421-31.

Brown, G. W., Ed. 1968, 1974. *Desert biology*, vols. 1 and 2. New York: Academic Press.

Brown, H. A. 1976. The status of California and Arizona populations of the western spadefoot toads (genus *Scaphiopus*). *Not. Hist. Mus. Los Angeles County Contrib. Sci.* 286: 1-15.

Brown, J. H. 1973. Species diversity of seed-eating desert rodents in sand dune habitats. *Ecology* 54: 775-87.

Brown, J. H., and Davidson, D. W. 1977. Competition between seed-eating rodents and ants in desert ecosystems. *Science* 196: 880-82.

Brown, J. H.; Davidson, D. W.; and Reichman, O. J. 1979. An experimental study of competition between seed-eating desert rodents and ants. *Amer. Zool.* 19: 1129-43.

Brown, J. H., and Lieberman, G. A. 1973. Resource utilization and coexistence of seed-eating desert rodents in sand dune habitats. *Ecology* 54: 788-97.

Brown, J. H.; Lieberman, G. A.; and Deugler, W. F. 1972. Woodrats and cholla: dependence of a small mammal population on the density of cacti. *Ecology* 53: 310-13.

Brown, J. H.; Reichman, O. J.; and Davidson, D. W. 1979. Granivory in desert ecosystems. *Ann. Rev. Ecol. Syst.* 10: 201-27.

Brown, M. F. Unpubl. Some aspects of the ecology of the desert isopod *Hemilepistus reaumuri* (Crustacea, Isopoda, Porcellionidae) in a first-order drainage basin in relation to available sediment formation. (Private circulation)

Brunhuber, B. S. 1970. Egg laying, maternal care and development of young in the scolopendromorph centipede *Cormocephalus anceps anceps* Porat. *Zool. J. Linn. Soc.* 49: 225-34.

Bursell, E. 1955. The transpiration of terrestrial isopods. *J. exp. Biol.* 32: 238-55.

Byzova, J. B. 1967. Respiratory metabolism in some millipedes (Diplopoda). *Rev. Ecol. Biol. Sol* 4: 611-24.

Chernov, J. I.; Striganova, B. R.; and Ananjeva, S. I. 1977. Soil fauna of the polar desert at Cape Cheluskin, Taimyr Peninsula, U.S.S.R. *Oikos* 29: 175-79.

Chew, R. M. 1965. Water metabolism of desert-inhabiting vertebrates. *Biol. Rev.* 36: 1-31.

———. 1977. Some ecological characteristics of the ants of a desert-shrub community in southeastern Arizona. *Amer. Midl. Nat.* 98: 33-41.

Chew, R. M., and Chew, A. E. 1970. Energy relationships of the mammals of a desert shrub (*Larrea tridentata*) community. *Ecol. Monogr.* 40: 1-21.

Chew, R. M., and Dammann, A. E. 1961. Evaporative water loss of small vertebrates as measured with an infra-red analyzer. *Science* 133: 384-85.

Chew, R. M., and De Vita, J. 1980. Foraging characteristics of a desert ant assemblage: functional morphology and species separation. *J. Arid Environ.* 3: 75-83.

Christiansen, K. 1964. Bionomics of Collembola. *Ann. Rev. Ent.* 9: 147-78.

Claussen, D. L. 1969. Studies on water loss and rehydration in anurans. *Physiol. Zoöl.* 42: 1-14.

Cloudsley-Thompson, J. L. 1956. Studies in diurnal rhythms. VI. Bioclimatic observations in Tunisia and their significance in relation to the physiology of the fauna, especially woodlice, centipedes, scorpions and beetles. *Ann. Mag. nat. Hist.* 9: 305-29.

———. 1958. *Spiders, scorpions, centipedes and mites*. London and New York: Pergamon.

———. 1959. Studies in diurnal rhythms. IX. The water relations of some nocturnal tropical arthropods. *Ent. exp. appl.* 2: 248-56.

———. 1961a. Observations on the biology of the scorpion *Leiurus quinquestriatus* (H. & E.) in the Sudan. *Ent. mon. Mag.* 97: 153-55.

———. 1961b. Observations on the natural history of the 'camel-spider' *Galeodes arabs* C. L. Koch (Solifugae: Galeodidae) in the Sudan. *Ent. mon. Mag.* 97: 145-52.

———. 1961c. Some aspects of the physiology and behaviour of *Galeodes arabs*. *Ent. exp. appl.* 4: 257-63.

———. 1962. Lethal temperatures of some desert arthropods and the mechanism of heat death. *Ent. exp. appl.* 5: 270-80.

———. 1963a. Light responses and diurnal rhythms in desert Tenebrionidae. *Ent. exp. appl.* 6: 75-78.

———. 1963b. Some aspects of the physiology of *Buthotus minax* (Scorpiones: Buthidae) with remarks on other African scorpions. *Ent. mon. Mag.* 98: 243-46.

———. 1964. On the function of the sub-elytral cavity in desert Tenebrionidae (Col.). *Ent. mon. Mag.* 100: 148-51.

———. 1967. The water relations of scorpions and tarantulas from the Sonoran desert. *Ent. mon. Mag.* 103: 217-20.

———. 1969. Acclimation, water and temperature relations of the woodlice *Metaponorthus pruinosus* and *Periscyphus jannonei* in the Sudan. *J. Zool., Lond.* 158: 267-76.

———. 1975a. Adaptations of arthropods to desert environments. *Ann. Rev. Ent.* 20: 261-83.

———. 1975b. The desert as a habitat. In *Rodents in desert environments*, ed. I. Prakash, and P. K. Ghosh, pp. 1-13. Monogr. Biol. 28. The Hague: Junk.

Cloudsley-Thompson, J. L., and Chadwick, M. J. 1964. *Life in deserts*. London: Foulis.
Cloudsley-Thompson, J. L., and Crawford, C. S. 1970. Water and temperature relations, and diurnal rhythms of scolopendromorph centipedes. *Ent. exp. appl.* 13: 187-93.
Collins, M. S. 1969. Water relations in termites. In *Biology of Termites*, ed. K. Krishna and F. M. Weesner, 1: 433-58. New York and London: Academic Press.
Crawford, C. S. 1972. Water relations in a desert millipede *Orthoporus ornatus* (Girard) (Spirostreptidae). *Comp. Biochem. Physiol.* 42A: 521-35.
——. 1976. Feeding-season production in the desert millipede *Orthoporus ornatus* (Girard) (Diplopoda). *Oecologia (Berl.)* 24: 265-76.
——. 1978. Seasonal water balance in *Orthoporus ornatus*, a desert millipede. *Ecology* 59: 996-1004.
——. 1979a. Assimilation, respiration and production: (a) invertebrates. In *Arid-land ecosystems: structure, functioning and management*, ed. D. W. Goodall and R. A. Perry, 1: 717-29. International Biological Programme, vol. 16. Cambridge: Cambridge University Press.
——. 1979b. Desert detritivores: a review of life history patterns and trophic roles. *J. Arid Environ.* 2: 31-42.
Crawford, C. S., and Cloudsley-Thompson, J. L. 1971. Water relations and desiccation-avoiding behaviour in the vinegaroon *Mastigoproctus giganteus* (Arachnida: Uropygi). *Ent. exp. appl.* 14: 99-106.
Crawford, C. S., and Krehoff, R. C. 1975. Diel activity in sympatric populations of the scorpions *Centruroides sculpturatus* (Buthidae) and *Diplocentrus spitzeri* (Diplocentridae). *J. Arachnol.* 2: 195-204.
Crawford, C. S., and Matlack, M. C. 1979. Water relations of desert millipede larvae, larva-containing pellets, and surrounding soil. *Pedobiologia* 19: 48-55.
Crawford, C. S., and Riddle, W. A. 1974. Cold hardiness in centipedes and scorpions in New Mexico. *Oikos* 25: 86-92.
——. 1975. Overwintering physiology of the scorpion *Diplocentrus spitzeri*. *Physiol. Zoöl.* 48: 84-92.
Crawford, C. S.; Riddle, W. A.; and Pugach, S. 1975. Overwintering physiology of the centipede *Scolopendra polymorpha*. *Physiol. Zoöl.* 48: 290-94.
Crawford, C. S., and Wooten, R. C., Jr. 1973. Water relations of *Diplocentrus spitzeri*, a semi-montane scorpion from the southwestern United States. *Physiol. Zoöl.* 46: 218-29.
Creusere, F. M., and Whitford, W. G. 1976. Ecological relationships in a desert anuran community. *Herpetologica* 32: 7-18.
Crossley, D. A., Jr. 1977. Oribatid mites and nutrient cycling. In *Biology of oribatid mites*, ed. D. L. Dindal, pp. 71-85. Syracuse: State University of New York.
Crowe, J. H. 1971. Anhydrobiosis: an unsolved problem. *Amer. Nat.* 105: 563-73.
Crowe, J. H., and Madin, K. A. C. 1975. Anhydrobiosis in nematodes:

evaporative water loss and survival. *J. exp. Zool.* 193: 323-34.

Dantzler, W. H., and Schmidt-Nielsen, B. 1966. Excretion in fresh-water turtle (*Pseudemys scripta*) and desert tortoise (*Gopherus agassizii*). *Amer. J. Physiol.* 210: 198-210.

Davidson, D. W. 1977a. Foraging ecology and community organisation in desert seed-eating ants. *Ecology* 58: 725-37.

———. 1977b. Species diversity and community organisation in desert seed-eating ants. *Ecology* 58: 711-24.

Delson, J., and Whitford, W. G. 1973. Adaptations of the tiger salamander, *Ambystoma tigrinum*, to arid habitats. *Comp. Biochem. Physiol.* 46A: 631-38.

Délye, G. 1968. Recherches sur l'écologie, la physiologie et l'ethologie des Fourmis du Sahara. Ph.D. dissertation, Université d'Aix-Marseille.

———. (1969). Perméabilité du tégument et résistance aux temperatures elevées de quelques Arthropodes sahariens. *Bull. Soc. ent. Fr.* 75: 51-55.

Dimmitt, M. A., and Ruibal, R. 1980a. Environmental correlates of emergence in spadefoot toads (*Scaphiopus*). *J. Herp.* 14: 21-29.

———. 1980b. Exploitation of food resources by spadefoot toads (*Scaphiopus*). *Copeia* 1980: 854-62.

Dregne, H. E., ed. 1970. *Arid lands in transition*. Washington, D.C.: American Association for the Advancement of Science.

Dresco-Derouet, P. L. 1960. Le métabolism respiratoire des scorpions. I. Existence d'un rhythme nychéméral de la consommation d'oxygène. *Bull. Mus. Hist. Nat.* 32: 553-57.

———. 1964. Le métabolism respiratoire des scorpions. II. Mesures de l'intensité respiratorie chez quelques espèces à différentes températures. *Bull. Mus. Hist. Nat.* 36: 97-99.

Dunbar, B. S., and Winston, P. W. 1975. The site of active uptake of atmospheric water in larvae of *Tenebrio molitor*. *J. Ins. Physiol.* 21: 495-500.

Edney, E. B. 1958. The microclimate in which woodlice live. *Trans. Int. Congr. Ent.* 2: 709-12.

———. 1964a. Acclimation to temperature in terrestrial isopods. I. Lethal temperature. *Physiol. Zoöl.* 37: 364-377.

———. (1964b). Acclimation to temperature in terrestrial isopods. II. Heart rate and standard metabolic rate. *Physiol. Zoöl.* 37: 378-394.

———. (1966a). Absorption of water vapour from unsaturated air by *Arenivaga* sp. (Polyphagidae, Dictyoptera). *Comp. Biochem. Physiol.* 19: 387-408.

———. (1966b). Animals of the desert. In *Arid lands. A geographical appraisal*, ed. E. S. Hills, pp. 181-218. London & UNESCO Paris: Methuen.

———. (1968a). The effect of water loss on the haemolymph of *Arenivaga* sp. and *Periplaneta americana*. *Comp. Biochem. Physiol.* 25: 149-58.

———. (1968b). Transition from water to land in isopod crustaceans. *Am. Zool.* 8: 309-26.

———. 1971. Some aspects of water balance in tenebrionid beetles and a thysanuran from the Namib desert of southern Africa. *Physiol. Zoöl.* 44: 61-76.
Edney, E. B.; Haynes, S.; and Gibo, D. 1974. Distribution and activity of the desert cockroach *Arenivaga investigata* (Polyphagidae) in relation to microclimate. *Ecology* 55: 420-27.
Eisenberg, J. F. 1975. The behaviour patterns of desert rodents. In *Rodents in desert environments*, ed. I. Prakash and P. K. Ghosh, pp. 189-224. Monogr. Biol. 28. The Hague: Junk.
El-Ayouty, E. Y.; Ghabbour, S. I.; and El-Sayyed, A. M. 1978. Role of litter and the excreta of soil fauna in the nitrogen status of desert soils. *J. Arid. Environ.* 1: 145-55.
El-Kifl, A. H.; Wafa, A. K.; Shafiee, M. F.; and Shereef, G. M. 1970. Studies on land Isopoda in Giza region. *Bull. Soc. ent. Egypt* 54: 283-317.
Ettershank, G.; Ettershank, J. A.; and Whitford, W. G. 1980. Location of food sources by subterranean termites. *Environ. Entomol.* 9: 645-48.
Ettershank, G., and Whitford, W. G. 1973. Oxygen consumption of two species of *Pogonomyrmex* harvester ants (Hymenoptera: Formicidae). *Comp. Biochem. Physiol.* 46A: 605-11.
Evans, A. A. F., and Perry, R. N. 1976. Survival strategies in nematodes. In *The organisation of nematodes*, ed. N. A. Croll, pp. 382-423. London: Academic Press.
Ferrar, P., and Watson, J. A. L. 1970. Termites (Isoptera) associated with dung in Australia. *J. Aust. Ent. Soc.* 9: 100-102.
Fonteyn, P. J., and Mahall, B. E. 1979. Population dynamics of two co-dominating desert perennials: *Larrea tridentata* and *Ambrosia dumosa*. Paper presented to the American Institute of Biological Sciences meeting, Oklahoma State University, Stillwater, 12-17 August 1979.
Franco, P. J.; Edney, E. B.; and McBrayer, J. F. 1979. The distribution and abundance of soil arthropods in the northern Mojave desert. *J. Arid Environ.* 2: 137-49.
Freckman, D. W. 1978. Ecology of anhydrobiotic soil nematodes. In *Dry biological systems*, eds. J. H. Crowe and J. S. Clegg, pp. 345-57. New York: Academic Press.
Freckman, D. W.; Kaplan, D. T.; and Van Gundy, S. D. 1977. A comparison of techniques for extraction and study of anhydrobiotic nematodes from dry soils. *J. Nematol.* 9: 176-81.
Freckman, D. W., and Mankau, R. 1977. Distribution and trophic structure of nematodes in desert soils. *Ecol. Bull. (Stockh.)* 25: 511-14.
Freckman, D. W.; Mankau, R.; and Ferris, H. 1975. Nematode community structure in desert soils: nematode recovery. *J. Nematol.* 7: 343-46.
French, N. R. 1975. Activity patterns of a desert rodent. In *Rodents in desert environments*, ed. I. Prakash and P. K. Ghosh, pp. 225-239. Monogr. Biol. 28. The Hague: Junk.
French, N. R.; Stoddart, D. M.; and Bobek, B. 1975. Patterns of demography in small mammal populations. In *Small mammals, their produc-*

tivity and population dynamics, ed. F. B. Golley, K. Petrusewicz, and L. Ryszkowski, pp. 73-102. International Biological Programme, vol. 5. Cambridge: Cambridge University Press.

Fryer, G. 1957. Observations on some African millipedes. *Ann. Mag. nat. Hist.* 12: 47-52.

Fuller, W. H. 1974. Desert soils. In *Desert biology*, ed. G. W. Brown, Jr., pp. 31-101. New York: Academic Press.

Gaby, R. 1972. Comparative niche utilization by two species of kangaroo rats (genus *Dipodomys*). Ph.D. dissertation, New Mexico State University, 71 pp.

Ghabbour, S. I.; Mikhail, W. Z. A.; and Rizk, M. A. 1977. Ecology of soil fauna of Mediterranean desert ecosystems in Egypt. I. Summer populations of soil mesofauna associated with major shrubs in the littoral sand dunes. *Rev. Ecol. Biol. Sol* 14: 429-59.

Ghilarov, M. S. 1962. Termites of the USSR, their distribution and importance. In *Termites of the humid tropics*. Paris: UNESCO New Delhi Symposium.

Ghobrial, L. I., and Nour, T. A. 1975. The physiological adaptations of desert rodents. In *Rodents in desert environments*, ed. I. Prakash and P. K. Ghosh, pp. 413-44. Monogr. Biol. 28. The Hague: Junk.

Ghosh, P. K. 1975. Thermo-regulation and water economy in Indian desert rodents. In *Rodents in desert environments*, ed. I. Prakash and P. K. Ghosh, pp. 397-412. Monogr. Biol. 28. The Hague: Junk.

Gist, C. S., and Crossley, D. A., Jr. 1975. Feeding rates of some Cryptozoa as determined by isotopic half-life studies. *Environ. Ent.* 4: 625-31.

Grandjean, F. 1946. Les Enarthronota (Acariens). Première série. *Ann. Sci. Nat. Zool.* (11 ^{eme} série): 213-48.

Grassé, P. P. 1949. Ordre des Isoptères ou Termites. In *Traité de zoologie*, ed. P. P. Grassé, 9: 408-544. Paris: Masson et Cie.

Grassé, P. P., and Noirot, C. 1948. La 'climatisation' de la termitière par ses inhabitants et le transport de l'eau. *C. r. hebd. Séance Acad. sci. Paris* 227: 869-71.

Greenslade, P. In press. Survival of Collembola in arid environments: observations in South Australia and the Sudan. *Pedobiologia*.

Greenslade, P. J. M. 1979. *A guide to the ants of Australia*. Adelaide: South Australian Museum.

——— . In preparation. Three types of selection. (Private circulation)

Greenslade, P. J. M., and Greenslade, P. 1973. Epigaeic Collembola and their activity in a semi-arid locality in southern Australia during summer. *Pedobiologia* 13: 227-35.

——— . In preparation. Ecology of soil invertebrates.

Hadley, N. F. 1970a. Micrometeorology and energy exchange in two desert arthropods. *Ecology* 51: 434-44.

——— . 1970b. Water relations of the desert scorpion *Hadrurus arizonensis*. *J. Exp. Biol.* 53: 547-58.

——— . 1974. Adaptational biology of desert scorpions. *J. Arachnol.* 2: 11-23.

———. 1979. Thermal and water relations of desert animals. In *Arid-land ecosystems: structure, functioning and management*, ed. D. W. Goodall and R. A. Perry, 1: 743-68. International Biological Programme, vol. 16. Cambridge: Cambridge University Press.
Hadley, N. F., and Hill, R. D. 1969. Oxygen consumption of the scorpion *Centruroides sculpturatus*. *Comp. Biochem. Physiol.* 29: 217-26.
Hadley, N. F., and Williams, S. C. 1968. Surface activities of some North American scorpions in relation to feeding. *Ecology* 49: 726-34.
Hamilton, W. J., III. 1975. Coloration and its thermal consequences for diurnal desert insects. In *Environmental physiology of desert organisms*, ed. N. F. Hadley, pp. 67-89. Stroudsburg, Pa.: Dowden, Hutchinson and Ross.
Happold, D. C. D. 1975. The ecology of rodents in the northern Sudan. In *Rodents in desert environments*, ed. I. Prakash and P. K. Ghosh, pp. 15-45. Monogr. Biol. 28. The Hague: Junk.
Harris, W. V. 1970. Termites of the Palaearctic region. In *Biology of termites*, ed. K. Krishna and F. M. Weesner, 2: 295-313. New York and London: Academic Press.
Hawbecker, A. C. 1958. Survival and home range in the Nelson antelope ground squirrel. *J. Mammal.* 39: 207-15.
———. 1975. The biology of some desert-dwelling ground squirrels. In *Rodents in desert environments*, ed. I. Prakash and P. K. Ghosh, pp. 277-303. Monogr. Biol. 28. The Hague: Junk.
Hawke, S. D., and Farley, R. D. 1973. Ecology and behavior of the desert burrowing cockroach *Arenivaga* sp. (Dictyoptera, Polyphagidae). *Oecologia* 11: 263-79.
Hill, W. C. O.; Porter, A.; Bloom, R. T.; Seago, J.; and Southwick, M. D. 1957. Field and laboratory studies on the naked mole rat, *Heterocephalus glaber*. *Proc. zool. Soc. Lond.* 128: 455-514.
Hillyard, S. D. 1975. The role of anti-diuretic hormones in the water economy of the spadefoot toad, *Scaphiopus couchii*. *Physiol. Zoöl.* 48: 242-51.
Hoffman, R. L., and Payne, S. A. 1969. Diplopods as carnivores. *Ecology* 50: 1096-98.
Hohn, E., and Edney, E. B. 1973. Daily activity of Namib desert arthropods in relation to climate. *Ecology* 54: 45-56.
Hoover, K. D. 1973. Some ecological factors influencing the distributions of two species of pocket mice (genus *Perognathus*). Ph.D. dissertation, New Mexico State University.
Hoover, K. D.; Whitford, W. G.; and Flavill, P. 1977. Factors influencing the distributions of two species of *Perognathus*. *Ecology* 58: 877-84.
Horn, H. S. 1978. Optimal tactics of reproduction and life history. In *Behavioral ecology, an evolutionary approach*, ed. J. R. Krebs and N. B. Davies, pp. 411-29. Oxford: Blackwells.
Horne, F. R. 1973. The utilization of foodstuffs and urea production by a land snail during estivation. *Biol. Bull.* 144: 321-30.
Hudson, J. W.; Deavers, D. R.; and Bradley, S. R. 1972. A comparative

study of temperature regulation in ground squirrels with special reference to desert species. In *Comparative physiology of desert animals*, ed. G. M. O. Maloiy, pp. 191-213. Symp. zool. Soc. Lond., vol. 31. London: Academic Press.

Johnson, K. A., and Whitford, W. G. 1975. Foraging ecology and relative importance of subterranean termites in Chihuahuan desert ecosystems. *Environ. Ent.* 4: 66-70.

Kaestner, A. 1968. *Invertebrate zoology*, vol. 2. New York: Interscience.

Kassas, M. 1970. Desertification versus potential recovery in circumsaharen territories. In *Arid lands in transition*, ed. H. E. Dregne, pp. 123-42. Washington, D.C.: American Association for the Advancement of Science.

Kay, C. A. R., and Whitford, W. G. 1975. Influence of temperature and humidity on oxygen consumption of five Chihuahuan desert ants. *Comp. Biochem. Physiol.* 52A: 281-86.

———. 1978. Critical thermal limits of desert honey ants; possible ecological implications. *Physiol. Zoöl.* 51: 206-13.

Kay, F. R., and Whitford, W. G. 1978. The burrow environment of the bannertailed kangaroo rat, *Dipodomys spectabilis*, in south-central New Mexico. *Amer. Midl. Nat.* 99: 270-79.

Kheirallah, A. M. 1979. The population dynamics of *Periscyphis granai* (Crustacea: Isopoda) in the western highlands of Saudi Arabia. *J. Arid Environ.* 2: 329-37.

———. 1980. Aspects of the distribution and community structure of isopods in the Mediterranean coastal desert of Egypt. *J. Arid Environ.* 3:69-74.

———. In press. The life history and ecology of *Leptotrichus panzerii* (Crustacea: Isopoda) in Egypt. *Rev. Ecol. Biol. Sol.*

Kheirallah, A. M., and Awadallah, A. In press. The life history of the isopod *Porcellio olivieri* (And.) in the Mediterranean coastal desert of Egypt. *Pedobiologia*.

Köppen, W. 1954. Classification of climates and the world patterns. In *An introduction to climate*, ed. G. T. Trewartha, pp. 225-26, 381-83. 3d ed. New York: McGraw-Hill.

Kozlovskaja, L. S., and Striganova, B. R. 1972. Food, digestion and assimilation in desert woodlice and their relations to the soil microflora. In *Organisms as components of ecosystems*, ed. U. Lohm and T. Persson, pp. 240-45. Stockholm: Ecological Bulletin.

Kramm, R. A., and Kramm, K. R. 1972. Activities of certain species of *Eleodes* in relation to season, temperature and time of day at Joshua Tree National Monument (Coleoptera: Tenebrionidae). *Southwest. Nat.* 16: 341-55.

LaFage, J. P.; Haverty, M. I.; and Nutting, W. L. 1976. Environmental factors correlated with the foraging behavior of a desert subterranean termite *Gnathamitermes perplexus* (Banks) (Isoptera: Termitidae). *Sociobiology* 2: 155-69.

LaFage, J. P., and Nutting, W. L. 1978. Nutrient dynamics of termites. In

Production ecology of ants and termites, ed. M. V. Brian, pp. 165-232. International Biological Programme, vol. 13. Cambridge: Cambridge University Press.

Lawrence, R. F. 1953. *The biology of the cryptic fauna of forests.* Cape Town and Amsterdam: Balkema.

Lee, K. E., and Wood, T. G. 1971. *Termites and soils.* London and New York: Academic Press.

Lee, R. M. 1961. The variation of blood volume with age in the desert locust (*Schistocerca gregaria* Forsk.) *J. Insect Physiol.* 6: 36-51.

Lewis, J. G. E. 1971a. The life history and ecology of the millipede *Tymbodesmus falcatus* (Polydesmida: Gomphiodesmidae) in northern Nigeria with notes on *Sphenodesmus sheribongensis. J. Zool., Lond.* 164: 551-63.

———. 1971b. The life history and ecology of three paradoxosomatid millipedes (Diplopoda: Polydesmida) in northern Nigeria. *J. Zool., Lond.* 165: 431-52.

———. 1972. The life histories and distribution of the centipedes *Rhysida nuda togoensis* and *Ethmostigmus trigonopodus* (Scolopendromorpha: Scolopendridae) in Nigeria. *J. Zool., Lond.* 167: 399-414.

Linsemair, K. E. 1972. The importance of family-specific badges for the cohesion of families in the social desert woodlouse, *Hemilepistus reaumuri. Z. Tierpsychol.* 31: 131-62.

Lockard, R. B., and Owings, D. H. 1974. Seasonal variations in moonlight avoidance by Bannertail kangaroo rats. *J. Mammal.* 55: 189-93.

Loomis, H. F. 1966. Descriptions and records of Mexican Diplopoda. *Ann. ent. Soc Amer.* 59: 11-27.

Loots, G. C., and Ryke, P. A. J. 1967. The ratio Oribatei: Trombidiformes with reference to organic matter content in soils. *Pedobiologia* 7: 121-24.

Louw, G. N. 1972. The role of advective fog in the water economy of certain Namib desert animals. In *Comparative physiology of desert animals*, ed. G. M. O. Maloiy, pp. 297-314. Symp. zool. Soc. Lond., vol. 31. London: Academic Press.

Louw, G. N., and Hamilton, W. J., III. 1972. Physiological and behavioral ecology of the ultrapsammophilus Namib desert tenebrionid, *Lepidochora argentogrisea. Madoqua* 1: 87-95.

Ludwig, J. A., and Smith, S. D. 1978. Comparative primary production of Chihuahuan desert communities. *Bull. New Mexico Sci.* 18: 8.

MacArthur, R. H., and Wilson, E. O. 1967. *The theory of island biogeography.* Princeton, N.J.: Princeton University Press.

MacFarlane, W. V. 1975. Ecophysiology of water and energy in desert marsupials and rodents. In *Rodents in desert environments*, ed. I. Prakash and P. K. Ghosh, pp. 389-96. Monogr. Biol. 28. The Hague: Junk.

Machin, J. 1967. Structural adaptation for reducing water loss in three species of terrestrial snail. *J. Zool., Lond.* 151: 55-65.

———. 1975. Water balance in *Tenebrio molitor* L. larvae: the effect of

atmospheric water absorption. *J. Cell. Comp. Physiol.* 101: 121-32.
——. 1979. Atmospheric water absorption in arthropods. *Advances in Insect Physiol.* 14: 1-48.
Mackerras, M. J. 1970. Blattodea. In *Insects of Australia*, pp. 262—74. Commonwealth Scientific and Industrial Research Organization. Melbourne: Melbourne University Press.
MacMahon, J. A. 1979. North American deserts: their floral and faunal components. In *Arid-land ecosystems: structure, functioning and management*, ed. D. W. Goodall and R. A. Perry, 1: 21-82. International Biological Programme, vol. 16. Cambridge: Cambridge University Press.
MacMillen, R. E. 1972. Water economy of nocturnal desert rodents. In *Comparative physiology of desert animals*, ed. G. M. O. Maloiy, pp. 147-74. Symp. zool. Soc. Lond, vol. 31. London: Academic Press.
MacMillen, R. E., and Christopher, E. A. 1975. The water relations of two populations of non-captive desert rodents. In *Environmental physiology of desert organisms*, ed. N. F. Hadley, pp. 117-37. Stroudsburg, Pa.: Dowden, Hutchinson & Ross.
MacMillen, R. E., and Lee, A. K. 1967. Australian desert mice: independence of exogenous water. *Science* 158: 383-85.
——. 1969. Water metabolism of Australian hopping mice. *Comp. Biochem. Physiol.* 28: 493-514.
Madge, D. S. 1964. The water relations of *Belba geniculosa* Oudms. and other species of oribatid mites. *Acarologia* 6: 199-223.
Mares, M. A. 1975. Observations of Argentine desert rodent ecology, with emphasis on water relations of *Eligmodontia typus*. In *Rodents in desert environments*, ed. I. Prakash and P. K. Ghosh, pp. 155-75. Monogr. Biol. 28. The Hague: Junk.
Mares, M. A., and Rosenzweig, M. L. 1978. Granivory in North and South American deserts: rodents, birds and ants. *Ecology* 59: 235-41.
Martz, P. R. 1980. The foraging behaviour of a sand dune tenebrionid as a test of optimal foraging behaviour theory. Paper presented to the American Institute of Biological Sciences meeting, University of Arizona, Tucson, 3-8 August 1980.
Mason, C. F. 1970. Food, feeding rates and assimilation in woodland snails. *Oecologia (Berl.)* 4: 358-73.
Mayhew, W. W. 1968. Biology of desert amphibians and reptiles. In *Desert biology*, ed. G. W. Brown, pp. 195-356. New York: Academic Press.
Mazek-Fialla, K. 1934. Die lebensweise xerophiler Schnecken Syriens, Griechenlands, Dalmatiens und der Turkei und die Beschaffenheit ihrer subepithelialen Drüsen. *Z. Morph. Ökol. Tiere* 28: 445-68.
McBrayer, J. F. 1973. Exploitation of deciduous leaf litter by *Apheloria montana* (Diplopoda: Eurydesmidae). *Pedobiologia* 13: 90-98.
McClanahan, L. L., Jr. 1967. Adaptations of the spadefoot toad, *Scaphiopus couchi*, to desert environments. *Comp. Biochem. Physiol.* 20: 73-99.
——. 1972. Changes in body fluids of burrowed spadefoot toads as a

function of soil water potential. *Copeia* 1972: 209-16.
───. 1975. Nitrogen excretion in arid-adapted amphibians. In *Environmental physiology of desert organisms*, ed. N. F. Hadley, pp. 106-16. Stroudsburg, Pa.: Dowden, Hutchinson & Ross.
McGinnies, W. G. 1979a. General description of desert areas. In *Arid-land ecosystems: structure, functioning and management*, ed. D. W. Goodall and R. A. Perry, 1: 5-19. International Biological Programme, vol. 16. Cambridge: Cambridge University Press.
───. 1979b. Arid-land ecosystems—common features throughout the world. In *Arid-land ecosystems: structure, functioning and management*, ed. D. W. Goodall and R. A. Perry, 1: 299-316. International Biological Programme, vol. 16. Cambridge: Cambridge University Press.
McNab, B. K., and Morrison, P. 1963. Body temperature and metabolism in subspecies of *Peromyscus* from arid and mesic environments. *Ecol. Monogr.* 33: 63-82.
Meigs, P. 1953. World distribution of arid and semi-arid homoclimates. In *Review of research in arid zone hydrology*. Arid Zone Programme 1. Paris: UNESCO, pp. 203-9.
Misonne, X. 1959. Analyse zoogéographique des Mammifères de l'Iran. *Mem. Inst. r. Sci. nat. Belg. (2mè sér.)* 59: 1-157.
───. 1975. The rodents of the Iranian deserts. In *Rodents in desert environments*, ed. I. Prakash and P. K. Ghosh, pp. 47-58. Monogr. Biol. 28. The Hague: Junk.
Moeur, J. E., and Eriksen, C. H. 1972. Metabolic responses to temperature of a desert spider, *Lycosa (Pardosa) carolinensis* (Lycosidae). *Physiol. Zool.* 45: 290-301.
Moffett, D. F. 1975. Sodium and potassium transport across the isolated hindgut of the desert millipede *Orthoporus ornatus* (Girard). *Comp. Biochem. Physiol.* 50A: 57-63.
Moore, B. P. 1969. Biochemical studies in termites. In *Biology of termites*, ed. K. Krishna and F. M. Weesner, 1: 407-32. New York and London: Academic Press.
Morton, S. R. 1979. Diversity of desert-dwelling mammals: a comparison of Australia and North America. *J. Mammal.* 60: 253-63.
Murphy, P. W., ed. 1962. *Progress in soil zoology*. International Society of Soil Science. Soil Zoology Colloquium (1958). London: Butterworths.
Naumov, N. P. 1975. The role of rodents in ecosystems of the northern deserts of Eurasia. In *Small mammals, their productivity and population dynamics*, ed. F. B. Golley, K. Petrusewicz, and L. Ryszkowski, pp. 299-309. International Biological Programme, vol. 5. Cambridge: Cambridge University Press.
Nelson, J. F., and Chew, R. M. 1977. Factors affecting the seed reserves in the soil of a Mohave desert ecosystem, Rock Valley, Nye County, Nevada. *Amer. Midl. Nat.* 97: 300-20.
Nevo, E. 1979. Adaptive convergence and divergence of subterranean mammals. *Ann. Rev. Ecol. Syst.* 10: 269-308.

Newell, P. F., and Appleton, T. C. 1979. Aestivating snails—the physiology of water regulation in the mantle of the terrestrial pulmonate *Otala lactea*. *Malacologia* 18: 575-81.

Newell, P. F., and Machin, J. 1976. Water regulation in aestivating snails. *Cell Tiss. Res.* 173: 417-21.

Newsome, A. E., and Corbett, L. K. 1975. Outbreaks of rodents in semi-arid and arid Australia: causes, preventions, and evolutionary considerations. In *Rodents in desert environments*, ed. I. Prakash and P. K. Ghosh, pp. 117-53. Monogr. Biol. 28. The Hague: Junk.

Nichols, U. G. 1953. Food habits of the desert tortoise, *Gopherus agassizii*. *Herpetologica* 9: 65-69.

Nielsen, C. O. 1949. Studies on the soil microfauna II. The soil-inhabiting nematodes. *Natura jutl.* 2: 1-131.

Noble-Nesbitt, J. 1963. Transpiration in *Podura aquatica* L. (Collembola, Isotomidae) and the wetting properties of its cuticle. *J. exp. Biol.* 40: 681-700.

———. 1970a. Water balance in the firebrat, *Thermobia domestica* (Packard). *J. exp. Biol.* 52: 193-200.

———. 1970b. Water uptake from subsaturated atmospheres: its site in insects. *Nature* 225: 753-54.

Norris, K. S. 1967. Color adaptation in desert reptiles and its thermal relationships. In *Lizard ecology: a symposium*, ed. W. W. Milstead. Columbia: University of Missouri Press.

Noy-Meir, I. 1973. Desert ecosystems: environment and producers. *Ann. Rev. Ecol. Syst.* 4: 25-51.

Nunez, F. S., and Crawford, C. S. 1976. Digestive enzymes of the desert millipede *Orthoporus ornatus* (Girard) (Diploda: Spirostreptidae). *Comp. Biochem. Physiol.* 55A: 141-45.

———. 1977. Anatomy and histology of the alimentary tract of the desert millipede *Orthoporus ornatus* (Girard) (Diplopoda: Spirostreptidae). *J. Morph.* 151: 121-30.

Nutting, W. L., and Haverty, M. I. 1976. Seasonal production of alates by five species of termites in an Arizona desert grassland. *Sociobiology* 2: 145-53.

Nutting, W. L.; Haverty, M. I.; and LaFage, J. P. 1975. Demography of termite colonies as related to various environmental factors: population dynamics and role in the detritus cycle. United States/International Biological Programme Desert Biome Research Memorandum, Logan, Utah.

O'Connor, F. B. 1967. The Enchytraeidae. In *Soil biology*, ed. N. A. Burges and F. Raw, pp. 213-57. London: Academic Press.

O'Donnell, M. J. 1977. Hypopharyngeal bladders and frontal glands: novel structures involved in water vapour absorption in the desert cockroach, *Arenivaga investigata*. *Amer. Zool.* 17: 234.

———. 1978. The site of water vapour absorption in *Arenivaga investigata*. In *Comparative physiology—water, ions and fluid mechanics*, ed. K. Schmidt-Nielsen, L. Bolis, and S. Maddrell, pp. 115-21. Cambridge: Cambridge University Press.

Paris, O. H. 1963. The ecology of *Armadillidium vulgare* (Isopoda: Oniscoidea) in California grassland: food, enemies and weather. *Ecol. Monogr.* 33: 1-22.
Paris, O. H., and Pitelka, F. A. 1962. Population characteristics of the terrestrial isopod *Armadillidium vulgare* in California grassland. *Ecology* 43: 229-48.
Petter, F. 1975. La diversité des Gerbillidés. In *Rodents in desert environments*, ed. I. Prakash and P. K. Ghosh, pp. 177-83. Monogr. Biol. 28. The Hague: Junk.
Pianka, E. R. 1970. On 'r' and 'K' selection. *Amer. Nat.* 104: 592-97.
Pisarski, B. 1978. Comparison of various biomes. In *Production ecology of ants and termites*, ed. M. V. Brian, pp. 326-31. International Biological Programme, vol. 13. Cambridge: Cambridge University Press.
Poinsot, N. 1968. Cas d'anhydrobiose chez le Collembola *Subisotoma variabilis* Gisin. *Rev. Ecol. Biol. Sol* 5: 585-86.
———. 1974. Comportment de certains Collemboles dans les biotopes xeriques mediterraneens: un nouveau cas d'anhydrobiose. *C.R. Acad. Sci. Paris* 278: 2213-15.
Polis, G. A. 1979. Prey and feeding phenology of the desert sand scorpion *Paruroctonus mesaensis* (Scorpionidae: Vaejovidae). *J. Zool., Lond.* 188: 333-46.
———. 1980. Seasonal patterns and age-specific variation in the surface activity of a population of desert scorpions in relation to environmental factors. *J. Anim. Ecol.* 49: 1-18.
Polis, G. A., and Farley, R. D. 1979. Characteristics and environmental determinants of natality, growth and maturity in a natural population of the desert scorpion *Paruroctonus mesaensis* (Scorpionida: Vaejovidae). *J. Zool., Lond.* 187: 517-42.
———. 1980. Population biology of a desert scorpion: survivorship, microhabitat and the evolution of life history strategy. *Ecology* 61: 620-29.
Prakash, I. 1975. The population ecology of the rodents of the Rajasthan desert, India. In *Rodents in desert environments*, ed. I. Prakash and P. K. Ghosh, pp. 75-116. Monogr. Biol. 28. The Hague: Junk.
Pugach, S., and Crawford, C. S. 1978. Seasonal changes in hemolymph amino acids, proteins and inorganic ions of a desert millipede *Orthoporus ornatus* (Girard) (Diplopoda: Spirostreptidae). *Can. J. Zool.* 56: 1460-65.
Reichle, D. E. 1968. Relation of body size to food intake, oxygen consumption and trace element metabolism of forest floor arthropods. *Ecology* 49: 538-52.
Reichman, O. J. 1977. Optimization of diets through food preferences by heteromyid rodents. *Ecology* 58: 454-57.
———. 1979. Desert granivore foraging and its impact on seed densities and distributions. *Ecology* 60: 1085-92.
Reichman, O. J., and Oberstein, D. 1977. Selection of seed distribution types by *Dipodomys merriami* and *Perognathus amplus*. *Ecology* 58: 636-43.
Reichman, O. J.; Prakash, I.; and Roig, V. 1979. Food selection and

consumption. In *Arid-land ecosystems: structure, functioning and management*, ed. D. W. Goodall and R. A. Perry, 1: 681-716. International Biological Programme, vol. 16. Cambridge: Cambridge University Press.

Riddle, W. A. 1975. Water relations and humidity-related metabolism of the desert snail *Rabdotus schiedeanus* (Pfeiffer) (Helicidae). *Comp. Biochem. Physiol.* 51A: 579-83.

――――. 1978. Respiratory physiology of the desert grassland scorpion *Paruroctonus utahensis*. *J. Arid Environ.* 1: 243-51.

Riddle, W. A.; Crawford, C. S.; and Zeitone, A. M. 1976. Patterns of haemolymph osmoregulation in three desert arthropods. *J. Comp. Physiol.* 112: 295-305.

Riddle, W. A., and Pugach, S. 1976. Cold hardiness in the scorpion *Paruroctonus aquilonalis*. *Cryobiology* 12: 248-53.

Riechert, S. E. 1974. The pattern of local web distribution in a desert spider: mechanisms and seasonal variation. *J. Anim. Ecol.* 43: 733-46.

――――. 1976. Web-site selection in the desert spider *Agelenopsis aperta*. *Oikos* 27: 311-15.

Riechert, S. E.; Reeder, W. G.; and Allen, T. 1973. Patterns of spider distribution (*Agelenopsis aperta* (Gertsch)) in desert grassland and recent lava bed habitats, south-central New Mexico. *J. Anim. Ecol.* 42: 19-35.

Riechert, S. E., and Tracy, C. R. 1975. Thermal balance and prey availability: bases for a model relating web-site characteristics to spider reproductive success. *Ecology* 56: 265-84.

Rodriguez, J. G.; Wade, C. F.; and Wells, C. N. 1962. Nematodes as a natural food for *Macrocheles muscaedomesticae* (Acarina, Macrochelidae), a predator of the housefly egg. *Ann. Ent. Soc. Amer.* 55: 507-11.

Rose, F. L. 1980. Turtles in arid and semi-arid regions. Paper presented at the American Institute of Biological Sciences meeting, University of Arizona, Tucson, 3-8 August 1980.

Rosenzweig, M. L. 1968. Net primary productivity of terrestrial communities: prediction from climatological data. *Amer. Nat.* 102: 67-73.

Rosenzweig, M. L.; Smigel, B.; and Kraft, A. 1975. Patterns of food, space and diversity. In *Rodents in desert environments*, ed. I. Prakash and P. K. Ghosh, pp. 241-67. Monogr. Biol. 28. The Hague: Junk.

Ruibal, R.; Tevis, L.; and Roig, V. 1969. The terrestrial ecology of the spadefoot toad *Scaphiopus hammondi*. *Copeia* 1969: 571-84.

Saito, S. 1967. Productivity of high and low density populations of *Japonaria laminata armigera* (Diplopoda) in a warm temperate forest ecosystem. *Res. Population Ecol.* 9: 153-66.

Santos, P. F.; DePree, E.; and Whitford, W. G. 1978. Spatial distribution of litter and microarthropods in a Chihuahuan desert ecosystem. *J. Arid Environ.* 1: 41-48.

Schaefer, D. A., and Whitford, W. G. 1981. Nutrient cycling by the subterranean termite *Gnathamitermes tubiformans* in a Chihuahuan desert ecosystem. *Oecologia*. 48: 277-83.

Schmidt-Nielsen, K. 1964. *Desert animals: physiological problems of heat and water.* Oxford: Clarendon Press.
―――. 1975. Desert rodents: physiological problems of desert life. In *Rodents in desert environments,* ed. I. Prakash and P. K. Ghosh, pp. 379-88. Monogr. Biol. 28. The Hague: Junk.
Schmidt-Nielsen, K., and Bentley, P. J. 1966. Desert tortoise *Gopherus agassizii*: cutaneous water loss. *Science* 154: 911.
Schmidt-Nielsen, K.; Taylor, C. R.; and Shkolnik, A. 1971. Desert snails: problems of heat, water and food. *J. Exp. Biol.* 55: 385-98.
―――. 1972. Desert snails: problems of survival. In *Comparative physiology of desert animals,* ed. G. M. O. Maloiy, pp. 1-13. Symp. zool. Soc. Lond., vol. 31. London: Academic Press.
Schmoller, R. R. 1970. Terrestrial desert arthropods: fauna and ecology. *Biologist* 52: 77-98.
Schneider, P. 1975. Beitrag zur Biologie der afganischen Wustenassel *Hemilepistus aphganicus* Borutzky 1958 (Isopoda, Oniscoidea). Aktivitats-verlauf. *Zool. Anzeiger* 195: 155-70.
Schroder, G. D. 1979. Foraging behaviour and home range utilization of the bannertail kangaroo rat (*Dipodomys spectabilis*). *Ecology* 60: 657-65.
Schroder, G. D., and Rosenzweig, M. L. 1975. Perturbation analysis of competition and overlap in habitat utilization between *Dipodomys ordii* and *Dipodomys merriami. Oecologia* 19: 9-28.
Schumacher, A., and Whitford, W. G. 1974. The foraging ecology of two species of Chihuahuan desert ants: *Formica perpilosa* and *Trachymyrmex smithi neomexicanus. Insectes Sociaux* 21: 317-30.
―――. 1976. Spatial and temporal variations in Chihuahuan desert ant faunas. *Southwest. Nat.* 21: 1-8.
Seely, M. 1978. The Namib dune desert: an unusual ecosystem. *J. Arid Environ.* 1: 117-28.
Seinhorst, J. W. 1962. Extraction methods for nematodes inhabiting soil. In *Progress in soil zoology,* ed. P. W. Murphy, pp. 243-56. London: Butterworths.
Seymour, R. S. 1973. Energy metabolism of dormant spadefoot toads (*Scaphiopus*). *Copeia* 1973: 435-45.
Shachak, M. 1975. Some aspects of the structure and function of a desert ecosystem, and its use in a teaching programme of a field studies centre. Ph.D. dissertation, The Hebrew University, Jerusalem. (In Hebrew, with English summary)
―――. 1980. Energy allocation and life history strategy of the desert isopod *H. reaumuri. Oecologia (Berl.)* 45: 404-13.
Shachak, M.; Chapman, E. A.; and Orr, Y. 1976. Some aspects of the ecology of the desert snail *Sphincterochila boissieri* in relation to water and energy flow. *Israel J. Med. Sci.* 12: 887-91.
Shachak, M.; Chapman, E. A.; and Steinberger, Y. 1976. Feeding, energy flow and soil turnover in the desert isopod *Hemilepistus reaumuri. Oecologia* 24: 57-69.
Shachak, M.; Orr, Y.; and Steinberger, Y. 1975. Field observation on the

natural history of *Sphincterochila* (*S.*) *zonata* (Bourguignat, 1853) (= *S. boissieri* Charpentier, 1847). *Argamon. Israel J. Malac.* 6: 20-46.

Shachak, M.; Steinberger, Y.; and Orr, Y. 1979. Phenology, activity and regulation of radiation load in the desert isopod *Hemilepistus reaumuri*. *Oecologia* (Berl.) 40: 133-40.

Shaffer, D. T., Jr., and Whitford, W. G. 1981. Behavioral responses of a predator, the round-tailed horned lizard, *Phrynosoma modestum*, and its prey, honey pot ants, *Myrmecocystus* spp. *Amer. Midl. Nat.* 105: 209-16.

Shaw, J., and Stobbart, R. H. 1972. The water balance and osmoregulatory physiology of the desert locust (*Schistocerca gregaria*) and other desert and xeric arthropods. In *Comparative physiology of desert animals*, ed. G. M. O. Maloiy, pp. 15-38. Symp. zool. Soc. Lond., vol. 31. London: Academic Press.

Shkolnik, A. 1966. Studies in the comparative biology of Israel's two species of spiny mice (Genus *Acomys*). Ph.D. dissertation, The Hebrew University, Jerusalem.

Shkolnik, A.; Taylor, R. C.; Finch, V.; and Borut, A. 1980. Why do Bedouins wear black robes in hot deserts? *Nature* 283: 373-75.

Shoemaker, V. H. 1980. Amphibians in arid environments. Paper presented at the American Institute of Biological Sciences meeting, University of Arizona, Tucson, 3-8 August 1980.

Shoemaker, V. H.; McClanahan, L. L., Jr.; and Ruibal, R. 1969. Seasonal changes in body fluids in a field population of spadefoot toads. *Copeia* 1969: 585-91.

Shoemaker, V. H.; Nagy, K. A.; and Costa, W. R. 1974. The consumption, utilization and modification of nutritional resources by the jack rabbit (*Lepus californicus*) in the Mojave desert. United States International Biological Programme Desert Biome Research Memorandum: 74-125.

Shorthouse, D. J. 1971. Studies on the biology and energetics of the scorpion *Urodacus yashenkoi*. Ph.D. dissertation, Australian National University.

Smith, H. D., and Jorgensen, C. D. 1975. Reproductive biology of North American desert rodents. In *Rodents in desert environments*, ed. I. Prakash and P. K. Ghosh, pp. 305-30. Monogr. Biol. 28. The Hague: Junk.

Soholt, L. F. 1973. Consumption of primary productivity by a population of kangaroo rats (*D. merriami*) in the Mojave desert. *Ecol. Monogr.* 43: 357-76.

Southwood, T. R. E. 1976. Bionomic strategies and population parameters. In *Theoretical ecology: principles and applications*, ed. R. M. May. Oxford: Blackwells.

Starker-Leopold, A. 1963. *The desert*. Time-Life International New York: Life Nature Library.

Starling, J. H. 1944. Ecological studies on the Pauropoda of the Duke Forest. *Ecol. Monogr.* 14: 291-310.

Stearns, S. C. 1976. Life-history tactics: a review of the ideas. *Quart. Rev. Biol.* 51: 3-47.
———. 1977. The evolution of life history traits. *Ann. Rev. Ecol. Syst.* 8: 145-71.
Steinberger, Y.; Grossman, S.; and Dubinsky, Z. 1978. Energy and nitrogen balances in the desert woodlouse *Hemilepistus reaumuri*. In Abstracts: 370. *IInd. Intern. Congr. Ecol.*
Stradling, D. J. 1978. Food and feeding habits of ants. In *Production ecology of ants and termites*, ed. M. V. Brian, pp. 81-106. International Biological Programme, vol. 13. Cambridge: Cambridge University Press.
Striganova, B. R. 1972. Effect of temperature on the feeding activity of *Sammatiulus kessleri* (Diplopoda). *Oikos* 23: 197-99.
Tevis, L. 1958. Interrelationships between the harvester ant *Veromessor pergandei* (Mayr) and some desert ephemerals. *Ecology* 39: 695-704.
Thornthwaite, C. W. 1948. An approach toward a rational classification of climate. *Geograph. Rev.* 38: 55-94.
Thorson, T. B., and Svihla, A. 1943. Correlation of habits of amphibians with their ability to survive the loss of body water. *Ecology* 24: 372-81.
Tilbrook, P. J. 1967. The terrestrial invertebrate fauna of the Maritime Antarctic. *Phil. Trans. R. Soc. B.* 252: 261-78.
Toye, S. A. 1966a. The effect of desiccation on the behaviour of three species of Nigerian millipedes: *Spirostreptus assiniensis*, *Oxydesmus* sp. and *Habrodesmus falx*. *Ent. exp. appl.* 9: 378-84.
———. 1966b. The reactions of three species of Nigerian millipedes (*Spirostreptus assiniensis*, *Oxydesmus* sp. and *Habrodesmus falx*) to light, humidity and temperature. *Ent. exp. appl.* 9: 468-84.
———. 1967. Observations on the biology of three species of Nigerian millipedes. *J. Zool., Lond.* 152: 67-78.
———. 1970. Some aspects of the biology of two common species of Nigerian scorpions *J. Zool., Lond.* 162: 1-9.
Travé, J. 1963. Écologie et biologie des Oribates (Acariens) saxicoles et arboricoles. *Vie et Milieu* (supp.) 14: 1-267.
Turkowski, F. J., and Reynolds, H. G. 1974. Annual nutrient and energy intake of the desert cottontail *Sylvilagus auduboni* under natural conditions. United States International Biological Programme Desert Biome Research Memorandum: 74-124.
Tyndale-Biscoe, H. 1973. *Life of marsupials*. London: Edward Arnold.
Uekert, D. N.; Bodine, M. C.; and Spears, B. M. 1976. Population density and biomass of the desert termite *Gnathamitermes tubiformans* (Isoptera: Termitidae) in a shortgrass prairie: relationship to temperature and moisture. *Ecology* 57: 1273-80.
Valiachmedov, B. V. 1981. Termites *Anacanthotermes ahngerianus* (Isoptera: Hodotermitidae) and their influence on takyr formation in southwestern Tadjikistan (Central Asia). *Pedobiologia* 21: 242-56.
Volz P. 1951. Untersuchungen über die Microfauna des Waldboden. *Zool. Jb. (Syst.)* 79: 514-66.

Vorhies, C. T. 1945. Water requirements of desert animals in the Southwest. *Ariz. Agr. Exp. Sta. Tech. Bull.*, p. 107.

Walker, R. F., and Whitford, W. G. 1970. Soil water capabilities in selected species of anurans. *Herpetologica* 26: 411-18.

Wallwork, J. A. 1960. Some Oribatei from Ghana. I. Sampling localities. II. Some members of the Enarthronota Grandj. *Acarologia* 2: 368-88.

———. 1970. *Ecology of soil animals.* London: McGraw-Hill.

———. 1972a. Distribution patterns and population dynamics of the microarthropods of a desert soil in southern California. *J. Anim. Ecol.* 41: 291-310.

———. 1972b. Mites and other microarthropods from the Joshua Tree National Monument, California. *J. Zool., Lond.* 168: 91-105.

———. 1975. Energetics of soil mites: the experimental approach. *Proc. 4th. Int. Congr. Acarology*, pp. 69-73. Budapest: Akademie Kiado, 1974.

———. 1976. *The distribution and diversity of soil fauna.* London: Academic Press.

———. 1980. Desert soil microarthropods: an 'r'-selected system. In *Soil biology as related to land use practices*, pp. 759-69. Proc. VII Int. Coll. Soil Zool. Washington, D.C.: Environmental Protection Agency.

———. In preparation. Preliminary observations on the microarthropod fauna in Chihuahuan desert soils.

Walter, H., and Stadelmann, E. 1974. A new approach to water relations of desert plants. In *Desert biology*, ed. G. W. Brown, pp. 213-310. New York: Academic Press.

Warburg, M. R. 1965a. The evaporative water loss of three isopods from semi-arid habitats in South Australia. *Crustaceana* 9: 302-8.

———. 1965b. The microclimate in the habitats of two isopod species in southern Arizona. *Amer. Midl. Nat.* 73: 363-75.

———. 1965c. Water relations and internal body temperatures of isopods from mesic and xeric habitats. *Physiol. Zoöl.* 38: 99-109.

———. 1968. Behavioural adaptations of terrestrial isopods. *Amer. Zool.* 8: 545-59.

———. 1972. Water economy and thermal balance of Israeli and Australian Amphibia from xeric habitats. In *Comparative physiology of desert animals*, ed. G. M. O. Maloiy, pp. 79-111. Symp. zool. Soc. Lond. vol. 31. London: Academic Press.

Warburg, M. R.; Rankevich, D.; and Chasanmus, K. 1978. Isopod species diversity and the community structure in mesic and xeric habitats of the Mediterranean region. *J. Arid Environ.* 1: 157-63.

Watson, J. A. L., and Gay, F. J. 1970. The role of grain-eating termites in the degradation of mulga ecosystems. *Search* 1: 43.

Watson, J. A. L.; Lendon, C.; and Low, B. S. 1973. Termites in mulga lands. *Tropical Grasslands* 7: 121-26.

Webster, D. B., and Webster, M. 1980. Morphological adaptations of the ear in the rodent family Heteromyidae. *Amer. Zool.* 20: 247-54.

Weesner, F. M. 1970. Termites of the Nearctic region. In *Biology of termites*, ed. K. Krishna and F. M. Weesner, pp. 477-525. New York and London: Academic Press.

Whitford, W. G. 1976a. Foraging behavior of Chihuahuan desert harvester ants. *Amer. Midl. Nat.* 95: 455-58.
―――. 1976b. Temporal fluctuations in density and diversity of desert rodent populations. *J. Mammal.* 57: 351-69.
―――. 1978a. Foraging by seed-harvesting ants. In *Production ecology of ants and termites*, ed. M. V. Brian, pp. 107-10. International Biological Programme, vol. 13. Cambridge: Cambridge University Press.
―――. 1978b. Foraging in seed harvester ants, *Pogonomyrmex* spp. *Ecology* 59: 185-89.
―――. 1978c. Structure and seasonal activity of Chihuahuan desert ant communities. *Insectes Sociaux* 25: 79-88.
Whitford, W. G., and Bryant, M. 1979. Behavior of a predator and its prey: the horned lizard (*Phrynosoma cornutum*) and harvester ants (*Pogonomyrmex* spp.). *Ecology* 60: 686-94.
Whitford, W. G., and Creusere, F. M. 1977. Seasonal and yearly fluctuations in Chihuahuan desert lizard communities. *Herpetologica* 33: 54-65.
Whitford, W. G.; DePree, D. J.; Hamilton, P.; and Ettershank, G. 1981. Foraging ecology of seed harvesting ants, *Pheidole* spp., in a Chihuahuan desert ecosystem. *Amer. Midl. Nat.* 105: 159-67.
Whitford, W. G.; Dick-Peddie, S.; Walters, D.; and Ludwig, J. A. 1978. Effects of shrub defoliation on grass cover and rodent species in a Chihuahuan desert ecosystem. *J. Arid. Environ.* 1: 237-42.
Whitford, W. G., and Ettershank, G. 1975. Factors affecting foraging activity in Chihuahuan desert harvester ants. *Environ. Entomol.* 4: 689-96.
Whitford, W. G.; Freckman, D. W.; Elkins, N. Z.; Parker, L. W.; Parmalee, R.; Phillips, J.; and Tucker, S. In press. Diurnal migration and responses to simulated rainfall in desert soil microarthropods and nematodes. *Soil Biol. Biochem.*
Whitford, W. G.; Kay, C. A.; and Schumacher, A. 1975. Water loss in Chihuahuan desert ants. *Physiol. Zoöl.* 48: 390-97.
Whitford, W. G.; Meentemeyer, V; Seastedt, T. R.; Cromack, K., Jr.; Crossley, D. A., Jr.; Santos, P. F.; Todd, R. L.; and Waide, J. B. 1981. Tests of the AET model of litter decomposition in deserts and a clear-cut forest. *Ecology*. 62: 275-77.
Whitford, W. G., and Meltzer, K. H. 1976. Changes in O_2 consumption, body water and lipid in burrowed desert juvenile anurans. *Herpetologica* 32: 23-25.
Whittaker, R. H. 1970. *Communities and ecosystems*. New York: Macmillan.
Williams, S. C. 1969. Birth activities of some North American scorpions. *Proc. Calif. Acad. Sci.* 37: 1-24.
Wise, D. H. 1979. An experimental study of inter-specific competition in a beetle community. Paper presented to the American Institute of Biological Sciences meeting, Oklahoma State University, 12-17 August 1979.
Wood, T. G. 1971. The distribution and abundance of *Folsomides deserti-*

cola Wood (Collembola: Isotomidae) and other microarthropods in arid and semi-arid soils in southern Australia, with a note on nematode populations. *Pedobiologia* 11: 446-68.

Wood, T. G., and Sands, W. A. 1978. The role of termites in ecosystems. In *Production ecology of ants and termites*, ed. M. V. Brian, pp. 245-92. International Biological Programme, vol. 13. Cambridge: Cambridge University Press.

Wooten, R. C., Jr., and Crawford, C. S. 1974. Respiratory metabolism of the desert millipede *Orthoporus ornatus* (Girard) (Diplopoda). *Oecologia (Berl.)*. 17: 179-86.

———. 1975. Food, ingestion rates and assimilation in the desert millipede *Orthoporus ornatus* (Girard) (Diplopoda). *Oecologia (Berl.)* 20: 231-36.

Wooten, R. C., Jr.; Crawford, C. S.; and Riddle, W. A. 1975. Behavioural thermoregulation of *Orthoporus ornatus* (Diplopoda: Spirostreptidae) in three desert habitats. *Zool. J. Linn. Soc.* 57: 59-74.

Yeates, G. W. 1971. Feeding types and feeding groups in plant and soil nematodes. *Pedobiologia* 11: 173-79.

Yom-Tov, Y. 1970. Investigations on the ecology and survival of two snails in the Negev desert. Ph.D. dissertation, Tel Aviv University.

———. 1972. Field experiments on the effect of population density and slope direction on the reproduction of the desert snail *Trochoidea (Xerocrassa) seetzeni*. *J. Anim. Ecol.* 41: 17-22.

Zachariassen, K. E., and Hammel. H. T. 1976. Freeze-tolerance in adult tenebrionid beetles. *Norw. J. Zool.* 24: 349-52.

Name Index

Ahearn, G. A. 135, 136, 146, 158-60, 238, 243
Allen, T. 39
Ananjeva, S. I. 87
Anderson, J. M. 195
Appleton, T. C. 154
Awadallah, A. 51, 151
Ayyad, M. A. 54

Baker, G. H. 42
Banerjee, B. 41
Bar, Z. 52, 200
Bartholomew, G. A. 186
Bentley, P. J. 164
Billingham, J. 237
Block, W. 243-45
Bloom, R. T. 113
Blower, J. G. 91
Bobek, B. 80, 81, 113, 114
Bodine, M. C. 58
Borut, A. 183
Bouillon, A. 59, 155
Bradley, S. R. 131, 186
Brian, M. V. 56
Brody, A. R. 106
Brown, G. W. 21
Brown, H. A. 67
Brown, J. H. 63, 64, 74, 77-79, 82, 178, 204, 207, 208-10, 212, 213, 215
Brown, M. F. 47
Brunhuber, B. S. 31
Bryant, M. 71, 179, 199, 256, 258, 259
Bursell, E. 150

Chadwick, M. J. 38, 100, 128
Chapman, E. A. 50, 53, 154, 226, 237
Chernov, J. I. 87
Chew, A. E. 83
Chew, R. M. 62, 64, 83, 143, 162, 199, 204, 205
Christiansen, K. 93
Christopher, E. A. 164, 227, 248
Claussen, D. L. 67, 143
Cloudsley-Thompson, J. L. 18, 21, 29, 36, 38, 45, 92, 100, 128, 134, 135, 152, 159, 179, 188, 198, 255
Collins, M. S. 155
Corbett, L. K. 75, 81
Costa, W. R. 83
Crawford, C. S. 29, 33, 40-43, 50, 55, 56, 59, 61, 97, 126, 134, 135, 145-48, 173, 174, 179, 180, 188, 198, 226, 230, 233, 238, 241, 243, 244, 253
Craybill, 29
Creusere, F. M. 143, 255
Crumack, K., Jr. 175
Crossley, D. A., Jr. 102, 175
Crowe, J. H. 87

Dammann, A. E. 143
Dantzler, W. H. 165, 241
Davidson, D. W. 63, 64, 74, 77-79, 82, 204-8, 212, 213, 215
Deavers, D. R. 131, 186
Delson, J. 67
Délye, G. 135, 162
DePree, E. 27, 83, 102, 110, 111, 117, 205

283

Deugler, W. F. 178
De Vita, J. 199, 204, 205
Dimmitt, M. A. 67-69, 226
Dregne, H. E. 2
Dunbar, B. S. 156

Edney, E. B. 20, 55, 60, 150, 151, 156, 157, 185, 188, 200, 201, 241, 243
Eisenberg, J. F. 72, 194, 211, 212
El-Kifl, A. H. 51, 229
Elkins, N. Z. 174
Eriksen, C. H. 39, 137, 188, 243
Ettershank, J. A. 58
Ettershank, G. 58, 61, 63, 161, 177, 184, 205, 207
Evans, A. A. F. 87

Farley, R. D. 33, 34, 208, 234, 247, 248, 251, 253, 254
Ferrar, P. 58
Finch, V. 183
Flavill, P. 210
Fonteyn, P. J. 117
Freckman, D. W. 85, 87, 88, 166, 167, 174
French, N. R. 80, 81, 113, 114
Fryer, G. 42

Gaby, R. 209
Gay, F. J. 59
Ghabbour, S. I. 54, 55, 64, 92, 101, 204, 227, 237
Ghilarov, M. S. 128
Ghobrial, L. I. 131, 162, 178, 186
Ghosh, P. K. 177, 178, 180
Gibo, D. 55
Gist, C. S. 61, 102
Grandjean, F. 107-09
Grassé, P. P. 128
Greenslade, P. 63, 93-96, 101, 125, 166, 180
Greenslade, P. J. M. 63, 101, 166, 245, 261

Hadley, N. F. 33, 138, 153, 158, 159, 181-83, 222, 241, 243
Hamilton, P. 184, 205-08
Hamilton, W. J. III. 219-22
Hammel, H. T. 243
Happold, D. C. D. 77, 214
Harris, W. V. 99
Haverty, M. I. 57, 59
Hawbecker, A. C. 80, 211
Haynes, S. 55
Healey, I. N. 195
Hill, W. C. O. 113
Hillyard, S. D. 67, 141, 142

Hoffman, R. L. 42
Hoover, K. D. 210
Horn, H. S. 245
Horne, F. R. 154
Hudson, J. W. 131, 186

Johnson, K. A. 58, 59, 259
Jorgensen, C. D. 79-81, 212

Kaestner, A. 36, 42, 139
Kaplan, D. T. 88
Kassas, M. 2, 8
Kay, C. A. R. 161, 180, 184
Kay, F. R. 129
Kheirallah, A. M. 46, 51, 151, 229
Köppen, W. 2
Kraft, A. 212, 213
Krehoff, R. C. 33, 179, 198

LaFage, J. P. 56-58
Lawrence, R. F. 92
Lee, A. K. 164
Lee, K. E. 56, 59, 60, 99, 203
Lendon, C. 203
Lewis, J. G. E. 31, 40-44, 135
Lieberman, G. A. 178, 209
Linsemair, K. E. 47
Lockard, R. B. 80
Loots, G. C. 102
Louw, G. N. 70
Low, B. S. 203
Ludwig, J. A. 23

MacArthur, R. H. 245
McBrayer, J. F. 61
McClanahan, L. L. Jr. 67, 141-43
MacFarlane, W. V. 79
McGinnies, W. G. 2, 16, 21, 173
Machin, J. 52, 53, 154, 156
Mackerras, M. J. 54
MacMahon, J. A., 2, 10, 15
MacMillen, R. E. 164, 186, 227, 248
McNab, B. K. 187
Madge, D. S. 106
Madin, K. A. C. 87
Mahall, B. E. 117
Mankau, R. 85, 87
Mares, M. A. 75
Martz, P. R. 19, 61
Mayhew, W. W. 143, 144, 164, 165
Mazek-Fialla, K. 51
Meentemeyer, V. 175
Meigs, P. 2
Meltzer, K. H. 141, 143, 255
Mikhail, W. Z. A. 54, 55, 64, 92, 101, 205, 227, 237
Misonne, X. 178

Moeur, J. E. 39, 137, 188, 243
Moffett, D. F. 241
Moore, B. P. 155
Morrison, P. 187
Morton, S. R. 73, 74, 77, 211
Murphy, P. W. 166

Nagy, K. A. 83
Naumov, N. P. 77
Nevo, E. 113, 114, 194
Newell, P. F. 154, 155, 235
Newsome, A. E. 75, 81
Nichols, U. G. 71
Nielsen, C. O. 88
Noble-Nesbitt, J. 94, 156
Noirot, C. 128
Norris, K. S. 222
Nour, T. A. 131, 162, 178, 186
Noy-Meir, I. 119-21, 173-75
Nunez, F. S. 43, 226
Nutting, W. L. 56-59

Oberstein, D. 213
O'Connor, F. B. 88
O'Donnell, M. J. 156
Orr, Y. 47, 53, 130, 154, 234, 235
Owings, D. H. 80

Paris, O. H. 50, 152, 153
Parker, L. W. 174
Parmalee, R. 174
Payne, S. A. 42
Perry, R. N. 87
Petter, F. 163
Phillips, J. 174
Pianka, E. R. 245
Pisarski, B. 64
Pitelka, F. A. 50
Poinsot, N. 166
Polis, G. A. 33-35, 228, 233, 234, 247, 248, 251, 253-55
Porter, A. 113
Prakash, I. 71, 77, 83, 131, 208
Pugach, 145, 179, 241, 244

Reeder, W. G. 39
Reichman, O. J. 9, 63, 64, 71, 74, 77-79, 82, 83, 204, 207, 208, 212, 213, 215
Reynolds, H. G. 83
Riddle, W. A. 29, 52, 130, 145, 153, 173, 174, 179, 180, 241, 244
Riechert, S. E. 39
Rizk, M. A. 54, 55, 64, 92, 161, 205, 227, 237
Rodriguez, J. G. 111
Roig, V. 68, 71, 77, 83, 141
Rose, F. L. 72

Rosenzweig, M. L. 208, 212, 213
Ruibal, R. 67-69, 141, 142, 226
Ryke, P. A. J. 102

Sands, W. A. 58, 100
Santos, P. F. 27, 83, 85, 90, 91, 99, 102, 104, 110, 111, 117, 175
Schaefer, D. A. 56, 58
Schmidt-Nielsen, B. 165, 241
Schmidt-Nielsen, K. 52, 77, 80, 153, 154, 162-65, 178, 182-84, 216, 217, 239
Schneider, P. 50
Schroder, G. D. 80, 208, 213
Schumacher, A. 61, 63, 161, 184
Seago, J. 113
Seastedt, T. R. 175
Seely, M. 18, 20
Seinhorst, J. W. 88
Seymour, R. S. 67, 141
Shachak, M. 47, 50, 53, 126, 130, 149, 154, 177, 226, 229, 234, 235, 237, 251
Shaffer, D. T. Jr. 70, 199, 227, 256, 258
Shafiee, M. F. 51, 229
Shereef, G. M. 51, 229
Shkolnik, A. 52, 153, 154, 182-84
Shoemaker, V. H. 68, 83, 142
Shorthouse, D. J. 233
Smigel, B. 212
Smith, H. D. 79-81, 212
Smith, S. D. 23
Soholt, L. F. 9
Southwick, M. D. 113
Southwood, T. R. E. 245
Springett, J. 85, 91
Stadelmann, E. 2
Starling, J. H. 92
Stearns, S. C. 231, 245, 257
Steinberger, Y. 46, 47, 50, 53, 92, 130, 200, 226, 234, 235, 237, 251, 253, 258
Stinnett, K. 110
Stoddart, D. M. 80, 81, 113, 114
Stradling, D. H. 63
Striganova, B. R. 87
Svihla, A. 143

Taylor, C. R. 52, 153, 154, 182-84
Tevis, L. 68, 141
Thornthwaite, C. W. 2
Thorson, T. B. 143
Tilbrook, P. J. 93
Todd, R. L. 175
Toye, S. A. 33, 42, 44, 149
Tracy, C. R. 39
Travé, J. 167
Tucker, S. 174
Turkowski, F. J. 83
Tyndale-Biscoe, H. 66, 187, 229

Ueckert, D. N. 58

Valiachmedov, B. V. 9
Van Gundy, S. D. 88
Vorhies, C. T. 212

Wade, C. F. 111
Wafa, A. K. 51, 229
Waide, J. B. 175
Walker, R. F. 141
Wallwork, J. A. 12, 23, 26, 27, 39, 88, 92-94, 96, 97, 99, 101, 102, 104-07, 109-11, 124, 167, 168, 173, 176, 184, 188, 193, 195, 197, 232, 260
Walter, H. 2
Warburg, M. R. 45, 49, 67, 149-51, 229
Watson, J. A. L. 58, 59, 203
Webster, D. B. 80
Webster, M. 80
Weesner, F. M. 100
Wells, C. N. 111

Whitford, W. G. 27, 56, 58, 59, 61, 63, 64, 67, 70, 71, 83, 85, 90-92, 99, 102, 110, 111, 117, 129, 141, 143, 161, 162, 174, 175, 177, 179, 180, 184, 199, 205-08, 210, 211, 215, 227, 255, 256, 258, 259
Whittaker, R. H. 2
Williams, S. C. 33, 233
Wilson, E. O. 245
Winston, P. W. 156
Wise, D. H. 61
Wood, T. G. 56, 58, 59, 99, 100, 102, 203
Wooten, R. C., Jr. 41-43, 126, 180, 188, 243

Yeates, G. W. 85
Yom-Tov, Y. 51, 235, 251

Zachariassen, K. E. 243
Zeitone, A. M. 145, 241

Subject Index

Acacia greggii, 12
Acanthocyrtus, 94, 96
Acari (*see* mites)
Acaridae, 110
Acerontiella, 95, 96
Acomys, 73
Agelenopsis aperta, 39
Alaskozetes antarcticus, 167
Ambrosia (burr sage), 87; *dumosa*, 85
Ambystoma tigrinum, 67
Amitermes, 56
Amphibians, 36, 58, 65-69, 133, 140-43, 228, 243, 255-56
Amphibolorus, 70
Anacanthotermes, 128; *ahngerianus*, 9, 10; *ochraceus*, 56
Androctonus australis, 33, 137-40, 233, 250
Anhydrobiosis, 165-68; in Collembola, 95, 166, 167, 181, 232, 251; in mites, 102, 251; in nematodes, 87, 88, 95, 166, 167, 181, 232, 251, 259
Ants, 9, 56, 62, 63-64, 70, 81, 99-114, 126-29, 155, 168, 170, 177, 180, 189, 192, 199, 203, 204, 214, 224, 225, 232, 242, 247, 251, 258, 260; adaptive radiation in, 63-64; African, 64; Australian, 64; carnivorous, 199; densities of, 62, 258; diversity of, 62, 204; dormancy in, 29, 64, 181; ectothermism in, 64; feeding guilds in, 63, 192, 199; food storage by, 62, 63, 206; foraging by, 63, 74, 161, 177, 204, 206, 207; granivorous, 64, 192, 204, 214, 227, 238, 251 (*see also* harvester); harvester, 23, 28, 33, 35, 40, 56, 62, 74, 77, 126, 161, 171, 177, 184, 191, 199, 204, 214, 256, 258; honey-pot, 70, 161, 184, 191, 199, 227, 239, 256, 258 (*see also Myrmecocystus*); longevity in, 63-64; moisture requirements of, 161; nest-building by, 61; nocturnalism in, 161, 184, 224; North American, 64; predatory, 61, 63; sociality in, 64, 225, 258; surface activity in, 61, 62, 145, 168, 184; thermal limits of, 177, 184; water loss in, 135; wood-feeding, 63
Anurans, 67, 141-43, 226, 255
Apache plume, 14, 23
Aphelenchoides, 87
Aphelenchus, 87
Aporosaura anchietae, 70
Arachnids, 28-36, 85, 100-12, 133, 198, 228, 232, 240, 248, 249
Archistreptus, 40, 233, 253, 257
Arenivaga (sand roach), 26, 33, 54, 127, 135, 136, 155, 219, 241, 250; *investigata*, 158; osmoregulation by, 157; water uptake by, 155, 239
Arginase, 165
Armadillidae, 45, 49, 151-53
Armadillidium vulgare, 50, 150, 229, 230; oxygen consumption of, 188; thermal limits of, 185; transpiration rate of, 151-52
Armadillo, 250; *albomarginatus*, 45-51,

287

126, 150, 219; *officinalis*, 45-51, 150, 219, 229, 232, 236
Arroyo, 16, 23, 42, 110, 114, 115, 117, 206
Artemisia, 12
Astigmatid mites (Astigmata), 110, 124
Atriplex (saltbush), 10, 97-111, 125, 197

Bacteria, 27, 85, 86, 93, 101, 104
Bacteriophage, 93
Baiomys taylori, 79
Bajada, 12, 15, 21, 26, 62, 91, 92, 96, 125
Bandicoot, 66, 187
Bannertail kangaroo rat (*see Dipodomys spectabilis*)
Barchan dunes, 19, 60
Barking tree frog, 67
Bathyergidae, 112 (*see also* mole rats)
Bathyergus, 113
Bdellidae, 104, 196
Bedouin, 183
Billy gibber, 21
Birds, 8, 42, 58, 74, 203, 251
Bouteloua eriopoda, 99 (*see also* grama grass)
Brachystomella, 93, 96, 167
Buckwheat, 205
Bufonidae, 67
Bufo, 67; *boreus*, 67; *cognatus*, 67; *debilis*, 67; *marinus*, 143; *punctatus*, 67, 141; *regularis*, 67
Bulimulus dealbatus, 142, 153, 181-82
Burrowing owl, 29
Burrowing tree frog, 67
Buthidae, 31
Buthotus minax, 139
Buthus: *hottentotta*, 32, 126; *occitanus*, 34, 126, 138, 233, 248, 253

Caching, 9, 163, 191, 206, 227
Cacti, 77, 133, 162
Caeculidae, 104
Calcium oxalate, 72, 162
Caliche, 22, 26
Callisaurus, 70
Calosis, 200; *amabilis*, 131, 202
Camels, 186
Camel-spiders (*see* solifugids)
Campanotus, 63
Carabidae, 59, 220
Carapace, 133, 164
Cassia armata, 12
Cataglyphis, 63
Catclaw acacia (*see Acacia greggii*)
Centipedes, 29-31, 36, 42, 81, 91, 126, 129, 134, 174, 179, 182, 192, 218, 219, 224, 231, 232, 238, 247, 249, 254
Centrioptera muricata, 160

Centruroides sculpturatus, 25, 32, 126, 137, 139, 198, 214, 248, 250
Cepaea nemoralis, 220
Cephalobidae, 87
Ceratophrys, 68, 127, 142
Cercidium (*see* palo verde)
Cereus (*see* saguaro)
Chaparral, 212
Chelonians (*see* tortoises)
Chilopsis (*see* desert willow)
Cholla, 42, 127
Chott, 16
Chrysomelids, 97
Citellus, 78, 131, 162, 186, 211; *armatus*, 131, 186; *leucurus*, 131, 186, 211; *mohavensis*, 211; *nelsoni*, 80, 211; *spilosoma*, 131, 186; *tereticaudus*, 131, 186, 211; *townsendi*, 131, 186
Claypans, 12, 22, 23
Cnemidophorus, 70
Cockroach, 23, 36, 54, 127, 135, 136, 155, 249, 251, 255
Coelocnemis, 244
Coldenia, 10
Coleoptera, 54, 59-61; larvae, 155, 232
Collembola, 85, 91-95, 101, 125, 167-68, 180, 189-95, 251, 260; anhydrobiosis in, 95, 166, 167, 181, 232, 251; Australian, 93-95, 166; drought resistance in, 95, 166; feeding habits of, 93; life forms in, 94-96; Middle Eastern, 93-96; North American, 167
Conomyrma, 61, 227; *insana*, 199
Coptotermes, 100
Cormocephalus anceps, 31
Corynephoria, 94, 96, 166-67, 180
Cosmochthonius, 109
Cottontails, 82
Creosote bush, 12, 42, 58, 83, 96, 117, 144, 179, 188, 190, 197 (*see also Larrea tridentata*)
Cricetids, 81, 214, 227, 248, 254, 256
Crickets, 211
Cristataria, 172, 224
Crucifers, 119
Cryptocephalus, 61, 97
Cryptoglossa, 19, 26, 60, 220; *verrucosa*, 160
Cryptomys, 114
Cryptopygus, 93, 95
Cryptostigmatid mites (Cryptostigmata), 102-10, 124, 192, 194-97, 226, 231, 232, 238, 251, 260; breeding activity of, 124, 228; diet of, 109-10; hygrophilous, 105; mesophilous, 105; parthenogenesis in, 232; survival strategies in, 104-06; xerophilous, 105
Ctenolepisma terebrans, 99, 135, 156, 239

Ctenomyidae, 113 (see also tuco-tucos)
Ctenomys, 113, 128
Cunaxidae, 104
Cylindroiulus punctatus, 41

Dalea, 12
Damaeoidea, 105
Dasycercus cristicauda, 65, 144, 187
Dasylirion (see sotol)
Dasyuridae, 65, 87
Daya, 16
Dermestidae, 59
Desert oak, 14, 23, 125
Deserts: agriculture in, 2; animal husbandry in, 7, 8; Arabian, 4, 18, 46, 52, 170, 229; Argentine, 68, 75, 142; Atacama, 4, 6, 170, 173; Australian, 4, 54, 56, 65, 71-79, 90-95, 101, 102, 112, 142, 144, 173, 187, 202, 229, 230, 256; Chihuahuan, 8, 42, 56-61, 70, 92, 94, 99-112, 122, 145, 167, 168, 173, 179, 184, 194-99, 205, 232, 255; classification of, 2; coastal, 6, 18, 20; competition in, 74, 101, 110, 113, 189, 195-202, 205, 206, 209, 219, 251, 259; crop monoculture in, 7; definition of, 2; fossil water in, 7; global distribution of, 4, 6; Great Basin, 4, 6, 170, 173; habitat separation in, 29, 52; hot and cool, 4, 6, 119, 170-71, 175; human incursions in, 18; hydrology of, 10-12, 16, 23, 115-19; Indian, 170, 208; Iranian, 170; irrigation in, 7; Kalahari, 4, 71; litter accumulation in, 25, 91, 117, 149, 195-97, 206, 226; man's activities in, 7; microclimate in, 26-27, 174, 207; Middle Eastern, 40, 45, 51, 53, 92, 137, 153, 170, 208, 233, 237; Mojave, 7, 10, 19, 60, 85, 101-12, 117, 122, 166-68, 171-76, 204, 211, 228, 260; Namib, 6, 18, 20, 60, 70, 99, 130, 156, 160, 200, 202; Negev, 46, 53, 123, 170, 177, 183, 200, 235, 251; North American, 4, 18, 24, 29, 40, 45, 54, 67, 71-79, 82, 93, 104, 107, 112, 117-24, 126, 135, 139, 145, 150, 153-54, 159-60, 170-71, 179-81, 187, 194, 198, 206, 208-14, 228, 232, 233, 244, 247, 250; nutrient release in, 102-04; organic decomposition in, 102-04; overgrazing in, 7, 18; Patagonian, 4, 6, 171; productivity in, 2, 23, 27, 113; rainfall patterns in, 120-22, 138, 141, 169, 175, 225, 228, 237, 242; rain shadow, 6; reproductive strategies in, 124-25; resource allocation in, 29, 35, 53, 106-14; Saharan, 4, 8-19, 46, 56, 67, 112, 120, 128, 162, 170, 173, 214; soil erosion in, 7-15, 115; soils, 21-27, 192, 210, 232; Sonoran, 15, 39, 40, 58, 75, 104, 122, 137, 145, 160, 188, 204, 211, 229, 259; species diversity in, 189-97; species packing in, 31, 193, 259; subtropical, 4; Takla Makan/Gobi, 4, 6, 16, 170, 173; Thar, 4, 208; topography of, 10-15, 22, 115, 171; Turkestan, 4, 6, 18; wind action in, 16, 21, 26, 129, 206
Detritivore, 28, 46, 54, 59-64, 85, 87, 93, 99, 130, 144, 194-203, 226, 232, 234
Deuterosminthurus, 94, 180
Dew formation, 1, 55, 133, 143, 149, 163, 174, 226, 237
Dicranocarpus, 10
Dictyoptera, 54
Diet switching, 54, 63, 70, 77, 97, 133
Diplocentridae, 31
Diplocentrus spitzeri (= peloncillensis), 25, 33, 126, 137-39, 160-61, 174, 179, 198, 214, 244, 248, 250
Diplura, 85, 96
Dipodomys, 19, 77, 80, 161, 162, 189, 209, 214, 217, 224, 228, 239; agilis, 212; deserti, 79, 212; merriami, 79, 128, 164, 209, 212, 213, 248; microps, 79, 212; ordii, 79, 209; spectabilis, 79, 213
Dipsosaurus, 71; dorsalis, 144, 180, 189, 222
Diptera larvae, 97, 155, 194, 232
Dipus, 77
Ditylenchus, 85
Dormancy, 134, 145; in amphibians, 67, 68, 130, 255; in ants, 64, 181; in Collembola, 168, 181; in Hemilepistus, 124; in marsupials, 130; in millipedes, 124, 145; in nematodes, 87, 88, 167, 181 (see also anhydrobiosis); in rodents, 124, 131, 181; in sciurids, 80; in snails, 153, 183; in tenebrionids, 131, 181
Dormouse (see Eliomys melanurus)
Dorylaimina, 86
Dorymyrmex, 63
Drepanotermes perniger, 56
Dune formation, 10, 18
Dung: as food for beetles, 59; as food for termites, 58, 83, 100, 128, 203; as refuge for Thysanura, 97

Ecological vicariance, 73
Egernia kintorei, 144
Eleodes, 220, 244; armata, 158, 160; hispilabris, 241; obscurus, 61
Eleutherodactylus (see barking tree frog)
Eliomys melanurus, 235
Ellobius (see voles)
Elymus, 227
Enarthronota, 107

Enchytraeidae, 85, 91, 92, 255
Endogenous water, 132, 141, 142, 144, 161-65, 238-39
Entomobryidae, 94
Environmental stress, 2, 36, 41, 63-64, 66, 78, 122, 130, 138-48, 162, 165, 175, 180-81, 183-84, 185, 188, 201, 203, 216, 223, 238, 249, 256
Ephedra (*see* Mormon tea); *trifurca*, 179
Ereg, 16
Eremaeus, 105
Eremias, 71
Eremina: desertorum, 53; *ehrenbergi*, 237
Eremobates, 36
Erg, 16
Erioneuron pulchellum, 206
Erythraeidae, 104
Estivation, 181 (*see also* dormancy); in ants, 206, 207; in *Orthoporus*, 182; in rodents, 186, 207, 243; in snails, 142, 181, 182
Ethmostigmus trigonopodus, 31, 134
Eurypelma, 137, 160, 250
Euxerus erythropus, 214

Fallugia paradoxa, 92-99, 111 (*see also* Apache plume)
Flagellates, 203
Fluff grass, 205
Folsomides, 93-96, 166
Formica perpilosa, 63, 161, 184
Formicines, 63
Fouquieria splendens (*see* ocotillo)
Foxes, 80, 182
Franseria (*see* Ambrosia)
Funambulus pennanti, 180
Fungivore, 59, 86, 88, 99-110, 196

Galeodes arabs (= *granti*), 36, 136, 160, 198, 250, 255
Gallery, "carton", 58
Galumna, 106
Geomyidae, 112, 127, 194
Geomys, 128
Geophilomorpha, 91, 92
Geophilus, 92
Geotrupidae, 59
Gerbil, 77, 129, 131, 208, 214
Gerbillus, 73, 77, 81 (*see also* gerbil); *gleadowi*, 208; *nanus*, 208; *pyramidum*, 214; *watersi*, 214
Glycyphagus, 110, 124
Gnathamitermes, 56; *perplexus*, 58, 259; *tubiformans*, 56, 58
Gomphiodesmidae, 126, 256
Gopherus agassizii, 72, 164, 241
Grama grass, 96, 110, 197, 205

Granivores, 28, 63, 65, 74, 75, 133, 161-63, 198, 203, 206, 207, 208, 215, 227
Grasshopper mice (*see Onychomys*)
Ground squirrel, 78, 129, 131, 186, 211, 214, 249
Gryllids, 33
Gutteriezia sarothre (*see* snakeweed)
Gyrosis, 200; *moralesi*, 130-31, 202

Habrodesmus dubosqui, 42; *falx*, 149
Hadrurus arizonensis, 32, 137, 138, 153, 160, 181, 241, 243, 247, 250
Halophytes, 22, 163, 238
Hammada, 16
Haplochthonius, 107; *simplex*, 108, 232; *variabilis*, 107, 176, 228, 231, 232, 260
Haplozetes, 106
Hawks, 80
Helix aspersa, 154
Hemerotrecha californica, 35
Hemilepistus, 45, 46, 123-29, 151-53, 219, 231, 236, 248, 251; *afghanicus*, 49, 50; *reaumuri*, 229, 232, 237, 250, 257, 260; activity patterns in, 47-51, 123, 129, 149, 177; burrow system of, 46, 126, 225, 236; distribution of, 172; life history of, 46-51, 177-78; parental care in, 234, 236; photoreaction by, 49; sociality in, 49, 151, 225; transpiration rate of, 149-50, 152; water relations of, 149-50
Herbivore, 28, 46, 53, 54, 65, 69-79, 83, 87, 93, 114, 130, 133, 144-68, 194, 232, 234, 238
Heterocephalus glaber, 113
Heterogamia, 127, 237; *syriaca*, 55
Heteromyidae, 74, 80, 162, 210, 224, 230, 242, 248, 254, 256, 260
Hibernation (*see* dormancy)
Hilaria mutica (*see* Tabosa grass)
Holcomyrmex, 63
Homeostasis, 249
Honey-pot ants, 64, 70, 161, 180, 184, 199, 228, 239
Horned lizard (*see Phrynosoma*)
Hydrozetes, 105
Hygrophile, 105
Hylidae, 67
Hymenoptera, 54
Hypogastruridae, 95
Hypopus, 110, 124
Hystrix indica (*see* porcupine)

Ichnotropis, 71
Insectivora, 112
Insectivore, 65, 70, 75, 76, 85, 114, 130, 144, 162, 194, 227

Insects, 54-64, 66, 85, 96-100, 135, 145, 155-62, 193, 211, 232, 240, 250, 255, 258-59
Interference competition, 195, 205
Iridomyrmex, 199; *pruinosum*, 61
Ironwood (*Olneya*), 23
Isopods (*see* woodlice)
Isoptera, 56-59
Isotomidae, 95
Iteroparity: in *Orthoporus*, 233; in rodents, 252; in scorpions, 233, 252, 254; in snails, 235, 253; in social insects, 252; in tenebrionids, 252; in woodlice, 51

Jackrabbits, 82, 182
Jaculus, 73, 77, 161, 162; *jaculus*, 178, 214 (*see also* jerboa)
Japygidae, 96, 155, 193
Jasonia, 227
Jerboa, 178, 214
Joshuella, 105; *striata*, 124, 168, 176, 231, 260
Juniperus (juniper), 12, 15, 23, 92, 96-112, 124, 125, 176, 195, 197

Kalotermitidae, 155
Kangaroo, 77
Kangaroo rat, 19, 61, 72, 79, 127, 128, 189, 212, 213, 217, 228, 239 (*see also Dipodomys*)
Kavirs, 16
Kinosternon soriense, 72, 164
Krameria, 12, 87; *parvifolia*, 85
K-strategist (selection), 245-60

Lagomorphs, 9, 65, 82-83, 242 (*see also* jackrabbits, hares)
Larrea tridentata, 85, 87, 99, 101, 110 (*see also* creosote bush)
Lathriidae, 59
Leiurus quinquestriatus, 33, 137, 139, 161, 198, 234, 248, 250, 253
Lepidium lasiocarpa (*see* pepperweed)
Lepidochora, 200; *argentogrisea*, 130, 201; *porti*, 130, 201
Lepidocyrtus, 94, 96
Lepisma, 97
Leporillus, 77
Leptodactylidae, 67
Leptotrichus naupliensis, 51, 229
Lepus, 82
Lethal temperatures: in ants, 184; in snails, 184
Level-controling-flows, 119
Lichens, 167
Liposcelids, 99

Lithobius, 135
Lizards, 35, 58, 70, 129, 133, 134, 143, 144, 190, 198, 222, 224, 256, 258, 260; as food for marsupials, 65; as food for rodents, 211; diet switching in, 70-71, 76, 133, 256, 258; fossorial, 71, 144; granivory in, 71; temperature regulation in, 144, 179; uricotelism in, 144; water loss in, 143
Locusts, 9
Loess, 16, 22, 46, 52
Lotus, 119
Lycium andersoni, 85, 87
Lycosa carolinensis, 39, 137, 188, 243

Machilis, 97
Macropus robustus, 77
Macrotermes, 128
Macrotermitidae, 203, 228
Macrotis lagotis (*see* bandicoot)
Mallee, 56
Mammals, 85, 112-14; herbivorous, 114; insectivorous, 85, 114; subterranean, 112-14, 193
Marigold, 119
Marionina, 91
Marsupial mole, 65, 114 (*see also Notoryctes*)
Marsupial mouse (*see Sminthopsis*)
Marsupials, 65-66, 77, 112, 113, 187, 229, 243; fat storage in, 66, 187; herbivorous, 77; insectivorous, 130; litter size in, 66; nocturnal foraging by, 65, 187; subterranean, 65, 113; surface-active, 65, 134; temperature fluctuations in, 66, 251; torpor in, 66, 187
Mastigoproctus giganteus, 36, 136, 146, 160, 198, 238, 243, 250
Megaleia rufa, 77
Melophorus, 64
Meranoplus, 64
Meriones, 73, 77, 81, 228; (*see also* gerbil), *crassus*, 214; *hurrianae*, 129, 131, 177, 208
Mesa, 12
Mesophiles, 105
Mesostigmatid mites, 110, 125, 192, 197
Mesquite (*see Prosopis*)
Messor, 63, 204, 227
Metaponorthus, 51; *pruinosus*, 152, 185, 188, 229
Microdipodops, 77, 224; *megacephalus*, 212; *pallidus*, 212
Microzetes, 106
Millipedes, 23, 28, 39-44, 81, 91, 92, 119, 123-25, 129, 182, 188, 189, 192, 218, 219, 224-32, 237, 239, 240-43, 247-53,

256-57, 260; activity patterns in, 41, 42, 145, 180, 228; dormancy in, 40; feeding habits of, 42; life cycles of, 43-44

Mites, 27, 85, 92, 101-12, 124, 168, 193, 194, 259, 260; bacterial feeding, 85, 104, 109; feeding types in, 101; fungivorous, 104, 109; habitat preferences of, 195; nematophagous, 111; parthenogenesis in, 108; reproductive strategies in, 108, 124

Mole rats, 114, 128, 225

Moles, 114, 127, 194

Mollusks (see also snails), 28, 183, 192, 223, 234

Moloch horridus, 71

Monestrus, 66

Mormon tea, 42, 181

Morphospecies, 192

Muhlenbergia, 10

Mulgara, 66

Murids, 73, 75, 77, 81, 229, 230, 254, 256, 260

Mus, 81; *cervicolor*, 208; *musculus*, 210

Mutualism, 191

Myriapods, 85, 91-92

Myrmecia, 64

Myrmecines, 63

Myrmecocystus, 63, 70, 161, 184, 189, 199, 227, 256; *depilis*, 180, 199, 258; *mexicanus*, 61, 63, 184; *mimicus*, 180, 258; *romaini*, 180

Nama, 10

Nanorchestidae, 104

Nasal cooling, 164

Neanura, 93

Neanuridae, 95

Nebria brevicollis, 220

Neivamyrmex, 63

Nematalycidae, 104

Nematodes, 27, 85-88, 111, 165-68, 192, 194, 231, 240, 251, 259, 260; anhydrobiosis in, 87, 88, 166, 167, 232, 251; bacterial-feeding, 27, 85, 86, 101, 112; extraction of, 88, 165-66; feeding guilds in, 87; fungivorous, 86-87; omnivorous, 87; plant parasitic, 87; population densities of, 85, 87, 259; predatory, 87

Neobatrachus, 141

Neotoma, 72, 77, 79, 127, 162, 238; *lepida*, 162, 178, 212, 249

Nocturnalism, 126, 129, 134, 175, 226, 242, 249; in amphibians, 67, 140; in ants, 62-63, 161, 184, 224; in centipedes, 29, 42, 129, 224; in marsupials, 65, 144, 187; in rodents, 77-78, 129, 145, 162, 163, 211-13, 224; in scorpions, 32, 42, 129, 138, 181, 224, 233; in solifugids, 35, 129, 137, 224; in tenebrionids, 60-61, 130, 201, 224; in termites, 56, 224; in thysanurans, 99; in uropygids, 36, 129, 224; in woodlice, 46, 129, 144-45, 149, 151, 224

Notomys, 73, 77-81

Notoryctes, 113

Notoryctidae (see marsupial moles)

Novomessor, 56, 61, 63-64, 126, 162, 190, 227; *cockerelli*, 40, 63, 177, 184, 206

Ocotillo, 42

Octodont, 114, 225

Olneya (see ironwood)

Olpium kochi, 101

Ommatoiulus moreletii, 41

Omnivore, 64, 77, 87, 211

Oniscus asellus, 150, 185

Onychiuridae, 95

Onychiurus, 93, 96

Onychomys, 81, 228; *leucogaster*, 76, 79, 212; *torridus*, 76, 212, 249

Onymacris, 200; *bicolor*, 221; *caudidipennis*, 221; *laeviceps*, 201, 222; *plana*, 130, 160, 201, 202, 222; *rugatipennis*, 130, 202; *unguicularis*, 222

Oppiidae, 105, 110

Opuntia bigelovii (see cholla)

Oribatid mites, 167, 175, 176

Oribatula, 106

Orthoporus ornatus, 40-42, 126, 145-49, 188, 189, 225, 230, 237, 238, 241, 249, 250, 253, 257, 260; assimilation efficiency in, 43, 225; burrowing behavior in, 126; digestive enzymes of, 43; dormancy in, 145-48, 182; feeding activity in, 145, 180; iteroparity in, 233; longevity of, 43; metabolic compensation in, 188; molting in, 148; orientation behavior in, 180; spiral coiling in, 147, 219; transpiration rate of, 146-47; water relations of, 145-49

Otala lactea, 154, 155, 250

Othoes, 36

Owls, 80

Oxalic acid, 162

Oxidative metabolism, 133, 163

Oxydesmus, 149

Pack rat, 127, 178, 238 (see also *Neotoma*)

Palmatogecko, 71, 144

Palm squirrel, 179-80

Paloverde, 15, 23

Pandinus imperator, 139

Paradipus, 77
Paradoxosomatidae, 40-41, 42, 126, 256
Parental care, 231; in ants, 234; in centipedes, 31; in rodents, 80; in scorpions, 33, 228, 233, 234; in snails, 234; in solifugids, 234; in spiders, 234; in termites, 234; in woodlice, 47-48, 234, 236
Paruroctonus: *aquilonalis*, 241; *mesaensis*, 32-34, 228, 234, 248, 251, 254, 255
Passalozetes, 105-06, 168
Pauropoda, 91, 92, 125, 168, 193, 255, 259, 260
Pelobatidae, 67
Pepperweed, 15
Peramelidae, 66
Periscyphis: *granai*, 45-46, 229; *jannonei*, 152, 185, 188
Perognathus, 72, 77, 131, 190, 213; *amplus*, 213; *californicus*, 212; *fallax*, 248; *formosus*, 212; *intermedius*, 210; *longimembris*, 248; *penicillatus*, 210-14
Peromyscus, 77, 81, 210, 228; *crinitus*, 187, 227; *eremicus*, 79, 187, 212; *maniculatus*, 79, 212
Pheidole, 56, 61, 63-64, 184, 204, 215, 227; *desertorum*, 206; *gilvescens*, 206; *militicida*, 204, 206; *rugulosa*, 204; *sitarches*, 206; *xerophila*, 204, 206
Pheidologeton, 63
Pheromones, 240
Phrynosoma: *cornutum*, 70, 179, 199, 258; foraging behavior in, 70, 199; *modestum*, 70, 198, 227, 256, 258
Piñon pine, 12
Playa, 15-16, 69, 110, 132, 149, 206
Pleopods, 133, 150
Pocket gophers, 113-14, 128, 194 (*see also* Geomyidae)
Pocket mice, 127, 129, 131, 162, 190, 209, 213 (*see also Perognathus*)
Pogonomyrmex, 56, 61, 63-64, 71, 161, 177, 184, 199, 204, 215, 256, 259; *barbatus*, 206; *californicus*, 184, 259; *desertorum*, 63, 204, 259; *imberbiculus*, 204; *rugosus*, 63, 161, 184, 204, 259
Poisons, 240
Polar deserts, 1, 87, 93, 102, 167, 243
Polyestrous, 66, 228
Polyxenus, 91, 92
Ponerines, 63
Porcellio: *laevis*, 51, 185, 188; *olivieri*, 45-46, 51, 151, 229, 231, 253; photoreaction in, 151; *scaber*, 185
Porcellionidae, 45
Prorastriopes, 96, 167
Proformica, 64
Proisotoma, 93, 95; *brisbanensis*, 95

Pronuba, 191
Prosopis (mesquite), 10, 14, 16, 23, 42, 58, 77, 83, 92, 94, 96, 97, 101, 109, 111, 125, 167
Prostigmatid mites (Prostigmata), 102-104, 110-12, 124, 168, 192-97, 231, 251; anhydrobiosis in, 102, 251; predatory, 102, 104
Psammomys, 238; *obesus*, 162
Psammotermes, 128; *hybostoma*, 56
Pseudachorutes, 93
Pseudomys, 73, 77, 79, 81
Pseudoscorpions, 85, 100-01, 193, 255
Pseudosinella, 95
Pseudotracheae, 150, 153
Psocoptera, 97, 99, 155, 193, 194, 255, 259
Pternohyla (*see* burrowing tree frog)
Pterotermes occidentalis, 57
Pulse-reserve paradigm, 119-20, 203

Quercus grisea, 14, 23, 96, 97, 99, 125 (*see also* desert oak)

Rattlesnakes, 80
Rattus, 77, 81; *cutchicus*, 208
Reg, 16, 21
Reithrodontomys, 77, 81, 210, 228; *megalotis*, 79
Repletes, 63, 227
Reptiles, 42, 65, 69-72, 143-44, 164, 243
Repugnatorial secretions, 240; by arachnids, 134; by Collembola, 95; by millipedes, 42, 134, 148-49, 158, 249; by tenebrionids, 134, 158, 249; by uropygids, 36; by woodlice, 150
Respiratory quotient (RQ), 188
Rhizosphere, 85
Rhysida nuda togoensis, 31, 134
Rodents, 9, 72-82, 112, 119, 127-30, 133, 150, 162-63, 168, 177, 178, 186, 187, 203, 211-15, 223-25, 228, 230, 237, 240, 242, 248, 251-52, 256, 260; aggressive behavior in, 113, 210; auditory sensitivity in, 80; caching by, 78, 191; colony formation in, 114, 225; competition between, 113, 210-14; convergences among, 73, 113; crepuscular, 78, 187; diet switching, 77, 133, 212, 227; distribution of, 72-73, 112-13; diversity of, 73, 75, 208, 209, 214; dormancy in, 130, 181, 186; evolution of, 72, 74; feeding guilds in, 76, 192; feeding strategies of, 77, 162, 210; foraging by, 76, 77, 163, 214; granivorous, 23, 65, 72-82, 145, 162, 163, 191, 207-15, 224, 227, 228, 238, 251; habitat selec-

294 Index

tion by, 208-09; herbivorous, 77, 162, 211, 212; home range of, 114; insectivorous, 75-77, 211, 212; life expectancy in, 79; litter sizes in, 79-80; nest temperatures of, 177, 178; nocturnal, 77-78, 129, 130, 162, 163, 211, 212, 224; population densities of, 81, 256; reproductive biology of, 79, 80-81, 229, 252, 254; reproductive strategies in, 79-80, 125; resource allocation in, 208, 213; subterranean, 112, 127, 194, 210, 223, 225, 251; survival stragegies of, 81-82; temperature tolerance in, 186, 210; territoriality among, 114; thermoregulation in, 186, 187; water relations of, 162, 210
r-strategist (selection), 75, 245
Russian thistle, 42

Sage (see *Artemisia*)
Saguaro, 14
Salsola kali (see Russian thistle)
Saltbush, 10, 97, 101-02, 104, 109, 111, 125, 197 (see also *Atriplex*)
Sand dunes, 10, 12, 16, 18-19, 22-27
Sand rat (see *Meriones*)
Sauromalus, 71
Scaphiopus, 67-68, 127, 140-43, 145, 154, 226; *couchii*, 67-69, 141, 143, 226, 237; desiccation tolerance of, 141; diet of, 69; dormancy in, 68, 141, 142; emergence of, 68; fat storage in, 141; *hammondii*, 67, 141, 255; *holbrookii*, 67; juvenile mortality in, 142-43; *multiplicatus*, 68; ureotelism in, 141, 142
Scarabaeidae, 59
Sceloporus, 70
Scheloribates, 106
Sciurids, 80, 214, 254, 256
Scolopendra: *amazonica*, 31; *clavipes*, 134; *polymorpha*, 25, 29, 135, 174, 179, 241, 244, 250
Scolopendromorpha, 29, 126, 134-36, 138, 231, 232, 249, 252
Scorpio maurus, 126, 138, 140, 248, 251
Scorpionidae, 31
Scorpions, 25, 29, 31-34, 42, 70, 81, 100, 126-27, 129, 134, 136, 138-40, 158, 168, 174, 179, 181, 192, 198, 218, 224, 225, 232, 233, 238, 244, 247-54, 260; activity patterns in, 32; cannibalism, 34, 234, 255; feeding behavior in, 34; population dynamics of, reproduction and development in, 33-34, 228, 233, 252, 254; resource allocation in, 198; supercooling in, 244; transpiration rates of, 138, 161, 233; venoms of, 139-40, 158; viviparity in, 33; water conservation in, 137-40, 149
Scorpion-spiders, 32
Scutigeromorpha, 29
Scutovertex, 106
Sebkhas, 16
Seif dunes, 19
Seira, 94, 96
Semelparity: in insects, 232; in scorpions, 233, 252, 253; in woodlice, 51, 232, 252
Senna, 12
Serianus, 101
Serir, 16, 21
Setanodosa, 95, 167
Sierozems, 10, 22
Silphidae, 59
Silvilagus (see cottontail)
Skinks, 71, 144
Sminthopsis: *crassicaudata*, 65-66, 144, 187; *froggatti*, 65-66, 144, 187
Sminthuridae, 94-95, 167
Smoke tree, 12
Snails, 51-53, 119, 130, 145, 154, 155, 168, 174, 183, 190, 199, 218, 223, 224, 226, 232, 237, 240, 242, 247, 249, 251, 253, 257-58, 260; density-dependence in, 235; digestive enzymes of, 53; estivation in, 51, 53, 130, 154, 155, 181; feeding activities of, 53; longevity in, 53; reproductive biology of, 53, 228, 234; ureotelism in, 154; water relations of, 154, 155
Snakes, 69-70
Snake weed, 14
Solifugids, 31, 35-36, 100, 129, 134, 136, 160, 168, 198, 224, 250, 252, 255; cannibalism in, 35; diet of, 35; reproductive biology of, 35-36, 231
Solonchak, 22
Solonetz, 22
Solenopsis, 63-64; *xyloni*, 33
Sotol, 14
Spadefoot toads (see also *Scaphiopus*), 35, 67-68, 127, 140, 226; burrowing behavior in, 68; cocoon production by, 68, 142
Spalacidae, 112 (see also mole rats)
Spalacopus (see octodont)
Spalax, 128
Spermophilus, 78, 80; *mexicanus*, 80; *tereticaudus*, 80-81; *townsendi*, 81
Speyotyto cunnicularia (see burrowing owl)
Sphaeridia, 93, 125
Sphaerochthonius, 109
Sphenodesmus sheribongensis, 43

Sphincterochila, 52, 218, 221; *aharonii*, 52, 200; *cariosa*, 52, 200; *fimbriata*, 52, 200; *prophetarum*, 52-53, 130, 155, 200, 226, 235, 253; *zonata*, 51-53, 130, 153, 155, 182-84, 200, 221, 223, 225-26, 234-38, 242, 250, 258
Spiders, 31, 38-39, 70, 85, 100, 134, 182, 188, 193, 231, 232, 243; funnel-web, 39; tarantula, 38, 137, 182; trap-door, 38
Spinibdella cronini, 102, 110, 124, 168
Spirostreptidae, 40, 42
Springtails (*see* Collembola)
Staphylinidae, 59
Star dunes, 19
Stenocara phalangium, 222
Succulents, 133, 238
Suctobelbids, 110
Suaeda, 10 (*see also* iodinebush)
Sun-spider (*see* solifugids)
Symphypleona, 93

Tabosa grass, 15
Talpa, 113
Talpidae, 112 (*see also* moles)
Tapinoma, 63
Tarantula, 137, 160, 182
Tarsonemidae, 104, 195
Tatera, 81
Tectocepheus, 106, 168
Tegumental glands, 150
Tenebrio molitor, 156
Tenebrionid beetles, 20, 23, 59-60, 70, 83, 127, 129-31, 136, 145, 155, 158, 181, 182, 192, 198, 201, 211, 218-22, 232, 237, 240-44, 247-51, 255, 260; activity rhythms in, 60, 130, 159, 168, 200-02; coloration in, 182; crepuscular, 60, 200-02; feeding activity in, 60-61, 157, 202; habitat separation in, 200-02, 208; life histories of, 61; nocturnal, 60, 131, 224; water relations of, 60, 158, 159
Termites, 9-10, 28, 35, 54, 56-59, 70-71, 81, 83, 98-100, 125-26, 129, 177, 191, 199, 203, 224-27, 232, 242, 248, 250, 258; cannibalism by, 58; densities of, 56; distribution of, 56; drought tolerance in, 155; dry-wood, 155; feeding biology of, 56, 58, 100; foraging activities of, 56; harvester, 56; longevity of, 59; mound-building, 56, 59; nitrogen-fixation in, 58; nutrient recycling by, 58; reproductive biology of, 59; sociality in, 56, 225; swarming in, 59; trophylaxis in, 58
Termitidae, 100
Terrapene, 72; *horsfieldii*, 164
Testudines (*see* tortoises)

Thelyphonus caudatus, 36
Thermobia, 97; *domestica*, 156
Thrushes, 253
Thysanoptera, 97, 99, 155
Thysanura, 97, 99, 135, 155, 191, 232 (*see also* bristletails)
Ticks, 135, 156
Tiger salamander (*see Ambystoma tigrinum*)
Toads, 240
Torpor, 66, 151, 187, 222
Tortoises, 69, 71-72, 127, 129, 164, 168, 217; burrowing behavior in, 72, 164; diet of, 71-72, 164; fat storage in, 164; metabolic shifts in, 164; thermoregulation in, 186; water loss in, 164
Trachymyrmex smithi neomexicanus, 64
Transition temperatures: of *Hadrurus arizonensis*, 138; of *Orthoporus ornatus*, 146; of scolopendromorphs, 135; of tenebrionids, 160; of uropygids, 135, 146; of *Venezillo arizonicus*, 150
Transverse dunes, 19
Trochoidea seetzeni, 51-54, 153-54, 184, 218, 234-35, 253, 257
Trogidae, 59
Tuco-tucos, 113-14, 128, 225
Tufted millipede (*see Polyxenus*)
Tumulitermes tumuli, 203
Tydeidae, 103-4, 112
Tymbodesmus falcatus, 43
Typhlosaurus, 71

Ureotelism: in amphibians, 141-42; in snails, 154; in testudines, 165
Uricotelism, in reptiles, 69, 144, 165
Urodacus yaschenkoi, 233, 252, 253
Urodeles, 67
Uromastix, 71; *aegypticus*, 144; *hardwickii*, 144; *loricatus*, 144
Uropygids, 31, 36, 100, 129, 134-35, 168, 198, 224, 238, 243, 250, 252, 255
Urosaurus, 70
Uta stansburiana, 143

Vejovidae, 31-32
Venezillo arizonicus, 45-46, 126, 150, 152, 185, 219, 229, 230, 240, 250, 257
Venoms, 134, 139-40, 158
Verbena, 119
Veromessor, 63-64
Vinegaroon (*see Mastigoproctus*)
Voles, 128

Wadi, 16, 115, 151, 206
Whip-scorpions (*see* uropygids)
Wind scorpions (*see* solifugids)

Woodlice, 23, 28, 36, 45-51, 123, 126, 129, 133, 145, 149-53, 177, 192, 218, 219, 224, 225, 228-29, 237, 240-41, 243, 247-251, 253, 257, 260; conglobation in, 49, 126, 150, 151, 219, 250; evaporative cooling in, 45, 151; life cycles in, 50-51, 123; longevity in, 51; nocturnalism in, 46, 129, 145, 149, 151; temperature acclimation in, 185; thermal limits of, 185

Xanthine oxidase, 165
Xanthodesmus, 43; *physkon*, 42, 43

Xenolpium, 101
Xenylla, 93, 95, 166
Xeranianta veatchii, 53
Xerophiles, 105-6

Yucca, 12, 58, 191; *brevifolia*, 12 (*see also* Joshua tree); *elata*, 110, 179; *schidigera*, 12; soaptree, 14

Zygoribatula, 105
Zyzomys, 77